ERROR DETECTING CODES

General Theory
And Their Application In
Feedback Communication Systems

ERROR DETECTING CODES

General Theory And Their Application In Feedback Communication Systems

Torleiv KLØVE
Department of Informatics, University of Bergen
Bergen, Norway

Valery I. KORZHIK
St. Petersburg University of Telecommunications
St. Petersburg, Russia

SPRINGER SCIENCE+BUSINESS MEDIA, LLC

ISBN 978-1-4613-5976-0 ISBN 978-1-4615-2309-3 (eBook)
DOI 10.1007/978-1-4615-2309-3

Library of Congress Cataloging-in-Publication Data

A C.I.P. Catalogue record for this book is available
from the Library of Congress.

Printed on acid-free paper.

To Marit and Marina

CONTENTS

PREFACE

There are two basic methods of error control for communication, both involve coding of the messages. The differences lay in the way the codes are utilized. The codes used are block codes, which are the ones treated in this book, or convolutional codes. In most cases the block codes used are linear codes. With forward error correction, the codes are used to detect and correct errors. In a repeat request system, the codes are used to detect errors, and, if there are errors, request a retransmission.

Usually it is a much more complex task to correct errors than merely detect them. To detect errors has the same complexity as encoding, which is usually linear in the length of the codewords. Optimal error correcting decoding is in general an NP-hard problem, and efficient decoding algorithms are known only for some classes of codes. This has generated much research into finding new classes of codes with efficient decoding as well as new decoding algorithms for known classes of codes.

There are a number of books on error control, some are listed at the beginning of the bibliography at the end of this book. The main theme of these books is error correcting codes. Error detection tends to be looked upon as trivial and is covered in a few pages at most. What is then the reason behind the following book which is totally devoted to error detecting codes? The reason is, on the one hand, that error detection combined with repeat requests is the most widely used method of error control, and on the other hand, that the analysis of the reliability of the information transmission system with error detection is usually quite complex. Moreover, the methods of analysis are often not sufficiently known and simple rules of the thump are used instead.

Integrity of data is very important in modern communication systems. A significant property of error detecting codes is that they can, in contrast to forward error correcting codes, guarantee a required reliability of message transmission over any channel.

There are many papers devoted to the problem of the error detecting capability of a linear code, as well as for the binary symmetric channel as for other channels, with or without memory. They are spread over many journals and conference proceedings. The goal of this book has been to collect all these results and present them in a unified and clear form. We have also included a number of new results of our own, which have not been published before.

The book starts in Chapter 1 with a survey of the main channels models and in Chapter 2 we give a short introduction to the theory of block codes, with emphasis on the parts important for error detection. The weight distribution is particularly important for this application, and is treated in more detail than in most books on error correction.

In Chapter 3 we give a detailed account of the known results on the probability of undetected error on the important binary symmetric channel. Combined detection and correcting is also briefly mentioned, from the error detection point of view; decoding algorithms are not given. For the benefit of a reader who is mainly interested in how the various results may be utilized in search for an error detecting code for a particular application, we give a short guide, in Section 3.11, to help in such a search.

In Chapter 4 we discuss briefly error detection on some other channels, in particular the Z-channel and channels with memory. Further we give a description of the method of channel transforms which allow us to obtain high reliability even for trivial codes. This method can be used also on the BSC as a substitute, it will (at a cost) remove the need of "good" error detecting codes.

Error detecting codes are used in so-called automatic repeat request (ARQ) systems. In Chapter 5 we give an overview over the main ARQ protocols. To describe all such protocols would be beyond the scope of this book. We have concentrated on the main protocols in use and those which in our view are most promising for use in future systems. We have given a short description of a number of protocols in the comments at the end of the chapter. In Chapter 6 we discuss the optimal choice of codes for the various ARQ protocols.

Finally, we give a bibliography of papers on error detection and related subjects. When we have had a choice, we have given references to books and papers in English, but also a number of significant papers in Russian containing results, not found in English translation, are included.

Each chapter includes a list of comments and references.

The required background for the reader of this book is some basic knowledge of coding theory (typically an introductory university course on "coding theory" or "statistical communication theory") and some basic mathematical knowledge: algebra (matrices, groups, finite fields, vector spaces, polynomials) and probability theory (random events and variables, distribution functions, Markov chains).

Parts of material in the book has been tested in graduate courses in Norway and Russia.

The various parts of the book are of course written by the two authors based on their expertize. Chapters 1, 4.2, 4.3, 5, and 6 are mainly due to Korzhik, and chapters 2, 3, and 4.1 to Kløve. The reader will also recognize this in each authors individual style.

We hope that this book will be useful, both for coding theorists and mathematicians interested in error detection and for engineers employed with design of telecommunication systems, in particular with ARQ protocols. The probability of undetected error of linear codes has also found useful applications in frame synchronization, in the wire-tap channel, and in signature analysis for circuit testing. The book can also be used as a text book for a graduate course on error detection.

We thank Dr. V. Nikitin and Dr. P. Malev for doing the computations for some of the example and tables in Chapters 5 and 6, and also Dr. B. Kudryashov for useful discussion of his results on the FPMC model. Further, we thank a number of colleagues and students who have read and commented on parts of the manuscript for this book.

Bergen/St. Petersburg

1

CHANNEL MODELS

1.1 INTRODUCTION

A *channel model* is a description of the probability of receiving a block $\bar{y}$ when a block $\bar{x}$ is sent over the channel for all combinations of $\bar{x}$ and $\bar{y}$, we denote this by $P(\bar{y}|\bar{x})$. There are many different types of discrete channel models, but not all of them are necessary to describe encoding and decoding for feedback systems. The actual models have two main applications. The first is to calculate the ***probability of undetected error*** P_{ue} in some channel and the second to find the ***throughput efficiency*** for a given code, decoding algorithm, and protocol for the information system with feedback. On the other hand, the purpose of any channel model is to give a simplified description of a real channel without losing any of its main properties.

As will be shown in the next chapter, to find P_{ue} is not simple, even for the binary symmetric channel, which is one of the simplest models. For most codes one is not able to calculate P_{ue} for a complex model with memory or for a nonsymmetric channel. Fortunately. as will be shown in 4.2, there exists a method which guarantees that P_{ue} is below some value for any channel. The price one has to pay is a reduction in the throughput efficiency and increase in the complexity of encoding and decoding.

Therefore, we will restrict our description of channel models to models for which one can calculate P_{ue} or the throughput efficiency for powerful enough codes. An important value is the probability that a block of n symbols is received error free, $P_{\text{ef}}(n)$. An important function of a channel model is the ***multiplicity function***, which allows us to estimate P_{ue} for some code with known minimum distance.

We will restrict ourselves to channel models where one assumes that the received block has the same length as the sent block.

1.2 CHANNELS WITHOUT MEMORY

A channel has no memory if the errors are independent, that is

$$P\Big((y_1, y_2, \cdots, y_n) \;\Big|\; (x_1, x_2, \cdots, x_n)\Big) = \prod_{i=1}^{n} P(y_i|x_i).$$

The channel model is therefore completely described by the *transition probabilities* $P(y|x)$ for all x, y.

The q-ary Symmetric Channel

The q-ary *symmetric channel without memory*, qSC, is a discrete channel with q-ary input and output. The transition probabilities depend on a single parameter p as follows:

$$P(y|x) = \begin{cases} 1-p & \text{if } y = x, \\ p/(q-1) & \text{if } y \neq x. \end{cases} \tag{1.1}$$

In the particular case $q = 2$ we have the *binary symmetric channel without memory*, BSC .

The Error Multiplicity Function

The *error multiplicity function* $P(m, n)$ is the probability that there are at least m errors in a block of n symbols. For the symmetric channel qSC, $P(m, n)$ is given by

$$P(m,n) = \sum_{i=m}^{n} \binom{n}{i} p^i (1-p)^{n-i}. \tag{1.2}$$

It has the following bounds (see e.g. [16]).

$$\binom{n}{\lambda n} p^{\lambda n}(1-p)^{(1-\lambda)n} \le P(m,n) \le \frac{\lambda(1-p)}{\lambda - p}\binom{n}{\lambda n} p^{\lambda n}(1-p)^{(1-\lambda)n}, \tag{1.3}$$

where $\lambda = \frac{m}{n}$, and $\lambda < p$. If $\lambda > p$ then

$$P(m,n) \le \lambda^{-\lambda n}(1-\lambda)^{-(1-\lambda)n} p^{\lambda n}(1-p)^{(1-\lambda)n}. \tag{1.4}$$

For the qSC, the ***probability of an error free block*** of length n is clearly

$$P_{\mathrm{ef}}(n) = (1-p)^n. \tag{1.5}$$

The Binary Non-symmetric Channel

The *binary non-symmetric channel without memory* is given by

$$\begin{array}{ll} P(0|0) = 1 - p_1, & P(1|0) = p_1, \\ P(0|1) = p_2, & P(1|1) = 1 - p_2, \end{array}$$

where $p_1 \neq p_2$.

The most important non-symmetric channel is the *Z-channel* where $p_1 = p > 0$ and $p_2 = 0$.

1.3 CHANNELS WITH MEMORY

Let $\bar{x}$ and $\bar{y}$ be two q-ary vectors of length n. The corresponding *error pattern* of length n is the sequence

$$\bar{e} = (e_1 e_2 \cdots e_n) \in GF(2)^n$$

where $e_i = 0$ if $x_i = y_i$ and $e_i = 1$ otherwise.

A q-ary channel is a channel with ***additive noise*** if and only $P(\bar{y}|\bar{x})$ only depends on the corresponding error pattern:

$$P(\bar{y}|\bar{x}) = P(\bar{e}). \tag{1.6}$$

We restrict our attention of channels with memory to channels with additive noise. In particular, such channels are symmetric in the sense that $P(\bar{y}|\bar{x}) = P(\bar{x}|\bar{y})$ for all $\bar{x}, \bar{y}$.

The Varying Parameter Channel

A channel with additive noise is called a *varying parameter channel*, VPC, if the probability $P(e_i)$ of the i'th error pattern symbol e_i depends on a random parameter μ_i, and $0 \leq P(e_i) \leq 1/2$ for any μ_i.

The sequences $\mu_1, \mu_2, \cdots$ form a stochastic process with discrete time. It may have memory. The VPC is completely determined by $P(e_i|\mu_i)$ for all positive i and the joint probability density $w(\mu_1, \mu_2, \cdots, \mu_n)$ for all positive n. We note that it would be possible to let $\mu_i = P(e_i = 1)$, but it is more convenient in some cases to consider $P(e_i = 1|\mu_i)$ and μ_i separately. For a subset I of $\{1, 2, \cdots, n\}$, let e_I denote the error pattern with $e_i = 1$ if and only if $i \in I$. If I' is a subset of I, then obviously

$$P(e_I) \leq P(e_{I'}). \tag{1.7}$$

The Local Binomial Channel

The *local binomial channel* with *binomial interval* s is the VPC such that $\mu_{rs+1} = \mu_{rs+2} = \cdots = \mu_{rs+s} = \mu^{(r)}$ for $r \geq 0$. We denote it by LBC-s or just LBC. In particular the LPC-1 is the same as the VPC.

Clearly, on every symbol interval where $\mu_i = \mu^{(r)}$ the error pattern stream is described by a BSC with error probability $p^{(r)}$ and therefore (1.2) is true for that interval. The parameters $\mu^{(1)}, \mu^{(2)}, \cdots$ and consequently $p^{(1)}, p^{(2)}, \cdots$ can be random values with some given joint probability density $w(\mu^{(1)}, \mu^{(2)}, \cdots, \mu^{(n)})$ and $w(p^{(1)}, p^{(2)}, \cdots, p^{(n)})$ or they can be constants.

LBC can be used to model communication systems with diversity. For example a multi tone modem, where the parts of the code block are transmitted on different tones; a fading in the subchannels will result in different symbol error probability for every tone. Another example is a frequency hopping system, where the parts of the data block are transmitted on different frequencies with different signal-to-noise ratios.

The Composite Channel

The *composite channel*, CC, is the VPC where $\mu_i = \mu$ for all i. It will be completely described by $P(\mu) = P(e_i = 1|\mu_i)$ and the probability density function $w(\mu)$.

Let

$$\bar{P} = \int_\mu P(\mu)w(\mu)d\mu,$$

the average symbol error probability for this channel. Using Jensen's inequality we get

$$P_{\text{ef}} = \int_\mu (1 - P(\mu))^n w(\mu)d\mu \geq (1 - \bar{P})^n. \tag{1.8}$$

The Rayleigh Fading Channel

The *discrete mapping of the continuous Rayleigh fading channel*, or, for short, just *Rayleigh fading channel*, RFC, is the VPC where $P(e_i = 1|\mu_i) = P(\mu_i)$ is given by

$$P(\mu_i) = \frac{1}{2}\exp(-\tilde{h}^2\mu_i^2\alpha), \tag{1.9}$$

where

- $\tilde{h}^2 = E_s/N_0$, the signal-to-noise ratio at the input of the channel; E_s is the energy of a signal element, and N_0 is the spectral power density of the white Gaussian noise.
- α is a coefficient depending on the modulation and demodulation; for a binary frequency modulation (FM) and noncoherent receiving $\alpha = \frac{1}{2}$, for a binary phase shift keying (BPSK) and noncoherent demodulation $\alpha = 1$.

The random parameters μ_i are given by

$$\mu_i = \sqrt{\mu_{ci}^2 + \mu_{si}^2}, \tag{1.10}$$

where μ_{ci} and μ_{si} are Gaussian zero mean values with the same variances σ_μ^2, mutually independent, and having the the same correlation matrix $\mathbf{R}_\mu$.

The probability density function for RFC is the Rayleigh density function (hence its name):

$$w(\mu) = \begin{cases} \frac{\mu}{\sigma_\mu^2} \exp\left\{-\frac{\mu^2}{2\sigma_\mu^2}\right\}, & \text{for } \mu \geq 0 \\ 0 & \text{otherwise} \end{cases} \tag{1.11}$$

The average symbol error probability for RFC is given by the formula

$$\bar{P} = \frac{1}{2 + 2\alpha h^2}, \tag{1.12}$$

where $h^2 = 2\sigma_\mu^2 \tilde{h}^2$ is the signal-to-noise ratio at the output of the channel.

It is possible to obtain a useful, computable outcome for some choices of $\mathbf{R}_\mu$, e.g. when

$$\mathbf{R}_\mu = \begin{pmatrix} \sigma_\mu^2 & r & r^2 & \cdots & r^{n-2} & r^{n-1} \\ r & \sigma_\mu^2 & r & \cdots & r^{n-3} & r^{n-2} \\ \vdots & \vdots & \vdots & \ddots & \vdots & \vdots \\ r^{n-1} & r^{n-2} & r^{n-3} & \cdots & r & \sigma_\mu^2 \end{pmatrix}. \tag{1.13}$$

Let $P_0(i_1, i_2, \cdots, i_m)$ be the probability of having an error pattern with ones in positions $i_1, i_2, \cdots, i_m$. It has been shown [10, page 98] that this can be calculated as follows:

$$P_0(i_1, i_2, \cdots, i_m) =$$

$$det^{-1} \begin{pmatrix} A_1 & -2r^{i_2-i_1} & 0 & \cdots & 0 & 0 \\ -2^{i_2-i_1} & A_2 & -2r^{i_3-i_2} & \cdots & 0 & 0 \\ 0 & -2r^{i_3-i_2} & A_3 & \cdots & 0 & 0 \\ \vdots & \vdots & \vdots & \ddots & \vdots & \vdots \\ 0 & 0 & 0 & \cdots & -2r^{i_m-i_{m-1}} & A_m \end{pmatrix}, \tag{1.14}$$

where

$$\begin{aligned} A_k &= 2 + 2\alpha h^2 - 2\alpha h^2 r^{2(i_{k+1}-i_k)} + 2r^{2(i_{k+1}-i_k)} \text{ for } 1 \leq k \leq m-1, \\ A_m &= 2 + 2\alpha h^2. \end{aligned}$$

It is easy to show that the error pattern probabilities can be derived from (1.14). Let $I_m = \{i_1, i_2, \cdots, i_m\}$ and Z_k^n be the set of subsets of $\{1, 2, \cdots, n\}$ of size k.

Then

$$P(e_{I_m}) = \sum_{k=0}^{n-m}(-1)^k \sum_{\{j_1,j_2,\cdots,j_k\}\subseteq I_m} P_0(i_1,i_2,\cdots,i_m,j_1,j_2,\cdots,j_k),$$

$$P_{\text{ef}}(n) = 1+\sum_{k=1}^{n}(-1)^k \sum_{\{j_1,j_2,\cdots,j_k\}\in Z_k^n} P_0(j_1,j_2,\cdots,j_k).$$

A special case of (1.14) is the following:

$$P_0(1,2,\cdots,m) = \frac{1}{(2+2\alpha h^2)D_{m-1} - 4r^2 D_{m-2}}, \tag{1.15}$$

where D_m is found recursively as follows:

$$\begin{aligned} D_0 &= 1, \\ D_1 &= 2\alpha h^2(1-r^2) + 2(1+r^2), \\ D_m &= D_1 D_{m-1} - 4r^2 D_{m-2} \text{ for } m \geq 2. \end{aligned} \tag{1.16}$$

In particular, we get

$$P_0(1,2) = \frac{1}{4+8\alpha h^2 + 4\alpha^2 h^4(1-r^2)}, \tag{1.17}$$

$$P_0(1,2,3) = \frac{1}{8+24\alpha h^2 + 8\alpha^2 h^4(3-2r^2-r^4) + 8\alpha^3 h^6(1-r^2)^2}. \tag{1.18}$$

It has been shown that the following inequalities are valid:

$$P_0(e_{I_m}) \geq \frac{1}{(2+2\alpha h^2)^m}, \tag{1.19}$$

$$P_0(e_{I_m}) \leq \frac{1}{(2\alpha h^2)^m(1-r^2)^{m-1}}, \tag{1.20}$$

$$P(e_{I_m}) \leq P_0(e_{I_m}), \tag{1.21}$$

$$P_{\text{ef}}(n) \leq \sum_{i=0}^{n}(-1)^i \binom{n}{i} \frac{2^{n-i}}{R_{ni}}, \tag{1.22}$$

where

$$R_{ni} = \left(2 + 2\alpha h^2 \frac{i}{n}\right) D_{n-1,i} - 4r^2 D_{n-2,i},$$

and where

$$\begin{aligned} D_{0i} &= 1, \\ D_{1i} &= \frac{2\alpha h^2 i}{n}(1 - r^2) + 2(1 + r^2), \\ D_{ni} &= D_{1i} D_{n-1,i} - 4r^2 D_{n-2,i}. \end{aligned}$$

Remark. (1.19) is valid for any RFC, not only the exponential correlation case we consider here; and (1.21) is valid for any channel with additive noise.

Further

$$P_{\text{ef}}(n) \le U \stackrel{\text{def}}{=} \frac{2\alpha h^2 - n + 2}{2 + 2\alpha h^2} + \sum_{i=1}^{n} \frac{n - i}{4 + 8\alpha h^2 + 4\alpha^2 h^4 (1 - r^{2i})}, \tag{1.23}$$

$$P_{\text{ef}}(n) \ge U - \frac{\binom{n}{3}}{8 + 24\alpha h^2 + 4\alpha^2 h^4 (3 + 2r^2 + r^4) + 8\alpha^3 h^6 (1 - 2r^2 + r^4)}, \tag{1.24}$$

$$P(n, m) \le \sum_{i=m}^{n} \frac{\binom{n}{i}}{(2\alpha h^2)^i (1 - r^2)^{i-1}}. \tag{1.25}$$

For the particular case of a composite Rayleigh fading channel, CRFC, when $r = 1$ explicit recurrence relations for $P_{\text{ef}}(n)$ and $P(n, m)$ are known [10, page 93]:

$$P_{\text{ef}}(1) = \frac{2\alpha h^2 + 1}{2\alpha h^2 + 2}, \tag{1.26}$$

$$P_{\text{ef}}(n) = \frac{2\alpha h^2 n P_{\text{ef}}(n-1) + 2^{1-n}}{2\alpha h^2 n + 2}, \tag{1.27}$$

$$P(n,m) = \sum_{i=m}^{n} I(n,i), \tag{1.28}$$

where

$$I(m,m) = \frac{2^{1-m}}{2\alpha h^2 m + 2}, \tag{1.29}$$

$$I(s,m) = \frac{2\alpha h^2 s I(s-1,m) + 2^{1-s}\binom{s}{m}}{2\alpha h^2 s + 2}, \tag{1.30}$$

for $s = m, m+1, \ldots, n$.

It is easy to see that $P_{\text{ef}}(n)$ for CRFC decreases more slowly with n than $P_{\text{ef}}(n)$ for BSC. This is illustrated by the curve in Figure 1.1. This fact is very important for the throughput efficiency of feedback systems; see Chapter 5.

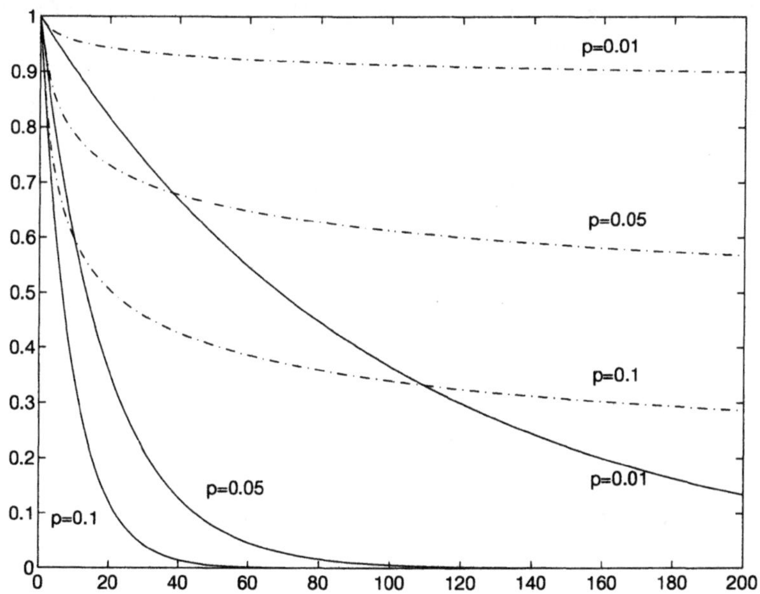

Figure 1.1 Plot of $P_{\text{ef}}(n)$ for the BSC (solid line) and CRFC with $\alpha = 1/2$ (dashed line) as a function of block length n for $p = 0.01$ ($h^2 = 98$), $p = 0.05$ ($h^2 = 18$), and $p = 0.1$ ($h^2 = 8$).

A Channel Model Based on a Function of a Pseudo-Markov Chain

From the relations for RFC one can calculate the error pattern probabilities or $P_{\mathrm{ef}}(n)$ for the CRFC efficiently, but not $P_{\mathrm{ef}}(n)$ or $P(n,m)$ for the RFC in the general case.

Therefore we consider the following interesting channel model which is capable to fill this gap in the theory. It is a *function of a pseudo-Markov chain*, FPMC.

Let $\mathbf{A}_0$ and $\mathbf{A}_1$ be two square $l \times l$ real matrices and let $\mathbf{A} = \mathbf{A}_0 + \mathbf{A}_1$. Further, let $\bar{a} = (a_1, a_2, \cdots, a_l)$ and $\bar{b} = (b_1, b_2, \cdots b_l)$ be two vectors of real numbers. We say that the error stream $\bar{e} = e_1, e_2, \cdots$ is a function of the corresponding pseudo-Markov chain if it satisfies

$$P(e_1, e_2, \cdots, e_n) = \bar{a}\mathbf{A}_{e_1}\mathbf{A}_{e_2} \cdots \mathbf{A}_{e_n}\bar{b}^{\mathrm{t}} \tag{1.31}$$

for all n. In particular, when $\mathbf{A}_0$ and $\mathbf{A}_1$ have only non-negative entries, $\mathbf{A}$ is stochastic, $\bar{a}$ is stochastic, and $\bar{b} = \bar{1}$, we say that $\bar{e}$ is a function of a Markov chain. The value l is called the *model rank*.

From (1.31) it is easy to derive the following expressions for the error multiplicity function:

$$P(n,s) = \sum_{m=s}^{n} \tilde{P}(n,m), \tag{1.32}$$

where

$$\tilde{P}(n,m) = \bar{a}(m,n)\bar{b}^{\mathrm{t}}, \tag{1.33}$$

where

$$\begin{aligned}
\bar{a}(-1,n) &= \bar{a} \text{ for } n \geq 1, \\
\bar{a}(0,1) &= \bar{a}\mathbf{A}_0, \\
\bar{a}(1,1) &= \bar{a}\mathbf{A}_1, \\
\bar{a}(m,n) &= \bar{a}(m-1,n-1)\mathbf{A}_1 + \bar{a}(m,n-1)\mathbf{A}_0.
\end{aligned}$$

The well known Gilbert-Elliot channel model is obtained as the following special case: $l = 2$ and

$$\mathbf{A}_0 = \begin{pmatrix} q_g P(g|g) & q_b P(b|g) \\ q_g P(g|b) & q_b P(b|b) \end{pmatrix}, \quad \mathbf{A}_1 = \begin{pmatrix} p_g P(g|g) & p_b P(b|g) \\ p_g P(g|b) & p_b P(b|b) \end{pmatrix},$$

where the states are either good (g) or bad (b), p_x is an initial probability to be in an x-state and $P(y|x)$ is the transition probability of going from an x-state to a y-state.

The notion of the FPMC model is very useful for the description of the block stream. In fact, the probability $P(N, M, n)$ that at least M out of N blocks are in error is given the formula obtained from (1.32) substituting $\mathbf{A}_0^n$ and $\mathbf{A}^n - \mathbf{A}_0^n$ for $\mathbf{A}_0$ and $\mathbf{A}_1$ respectively. In Chapter 5 we will use this approach to describe the throughput efficiency for channels with memory.

Another situation where there exists a description of the block stream is when $e_i = 0$ means that there are at most t errors in the i-th block of length n and $e_i = 1$ otherwise. The formula (1.32) remains true for this case if we change $\mathbf{A}_0$ to $\sum_{m=0}^{t} \mathbf{A}(m, n)$ and $\mathbf{A}_1$ to $\mathbf{A}^n - \mathbf{A}_0$, where $\mathbf{A}(m, n)$ can be found from the following recurrent matrix formula

$$\begin{aligned}
\mathbf{A}(0,1) &= \mathbf{A}_0, \\
\mathbf{A}(1,1) &= \mathbf{A}_1, \\
\mathbf{A}(m,n) &= \mathbf{A}(m,n-1)\mathbf{A}_0 + \mathbf{A}(m-1,n-1)\mathbf{A}_1.
\end{aligned}$$

It has been shown in [41] that if symbols or blocks are interleaved by s symbols or blocks respectively, then the previous formulas are true after changing $\mathbf{A}_i$ to $\mathbf{A}_i \mathbf{A}^{s-1}$ where $\mathbf{A} = \mathbf{A}_0 + \mathbf{A}_1$.

The FPMC model is very suitable for the cases when it is easy to find the probabilities of series of errors on the given channel. Let $P(1^i)$ denote the probability of having i successive ones in the error stream and let $P(1^i z 1^j)$ denote the probability of the sequence $1^i z 1^j$ in the error stream, where z is 0 or 1. Let

$\mathbf{Q}$ be the $l \times l$ matrix with elements $q_{ij} = P(1^{i+j-2})$,
$\mathbf{Q}_1$ be the $l \times l$ matrix with elements $q_{ij}^1 = P(1^{i+j-1})$,
$\mathbf{Q}_z$ be the $l \times l$ matrix with elements $q_{ij}^z = P(1^{i-1} z 1^{j-1})$.

It has been shown in [22] that

$$\begin{aligned}
&\mathbf{A}_0 = \mathbf{Q}^{-1}(\mathbf{Q}_z - \mathbf{Q}_1), \\
&\mathbf{A}_1 = \mathbf{Q}^{-1}\mathbf{Q}_1, \\
&\mathbf{A} = \mathbf{Q}^{-1}\mathbf{Q}_z,
\end{aligned} \tag{1.34}$$

$$\begin{aligned} \bar{a} &= (1, P(1), P(1^2), \cdots, P(1^{l-1})), \\ \bar{b} &= (1, 0, 0, \cdots, 0). \end{aligned}$$

For the RFC model, which is determined by the formulas (1.14)-(1.19) we obtain the following expressions for the entries of the matrices $\mathbf{Q}$, $\mathbf{Q}_1$, $\mathbf{Q}_2$:

$$P(1^i) = P_0(1, 2, \cdots, i), \text{ conf. (1.15)}, \tag{1.35}$$

$$P(1^i z 1^j) = S(i, j)^{-1}, \tag{1.36}$$

where

$$\begin{aligned} S(0, j) &= P(1^j), \\ S(1, j) &= c^* R_j - 4r^4 R_{j-1}, \\ S(i, j) &= cS(i-1, j) - 4r^2 S(i-2, j), \\ c &= 2\alpha h^2(1 - r^2) + 2(1 + r^2), \\ c^* &= 2\alpha h^2(1 - r^4) + 2(1 + r^4). \end{aligned}$$

Given the parameters h^2 and r for the RFC, it is easy to find $\mathbf{A}_0$, $\mathbf{A}_1$, $\mathbf{A}$, $\bar{a}$, $\bar{b}$ describing the FPMC model.

Remark. The description of some real error stream by the FPMC model is very useful to find its parameters by direct statistical testing.

Burst Error Channel Model (BEC)

A *burst* of errors of length i is an error pattern $\bar{e}$ which has all its ones among exactly i adjacent positions. We call a binary channel a BEC if the probability of an error pattern is given by

$$P(\bar{e}) = f(i, n), \tag{1.37}$$

where $i = i(\bar{e})$ is the length of the burst $\bar{e}$, and $f(i, n)$ is some function. There are situations where such a model is very close to reality. For example, if we have a wire channel with switching or a radio channel with pulse interferences, we obtain actually cut off channel for a presence of interferences or bad connections, and the probability of such noises depends upon their durations and consequently on the length of the burst.

1.4 CHANNEL TRANSFORMS

We can change the distribution of errors in a channel by putting the input and output of the channel through some transform. These transforms must not change the messages transmitted over the channel. They can be random (pseudorandom) or non-random. We will now describe some such transforms.

Outer Block Interleaving

The main example of a transform is the deterministic *outer block interleaving* or simply *symbol interleaving.* At the input of the channel (at the output of the encoder) we form the following sequence of new blocks:

$$\hat{b}_{ij} = b_{ji} \text{ for } 1 \leq i \leq s, \ 1 \leq j \leq n,$$

where

b_{ji} is the j-th symbol in the i-th old block (before transform),
$\hat{b}_{ij}$ is the i-th symbol in the j-th new block.

The integer s is usually called the *interleaving interval.* At the receiving end we perform the inverse transform. The procedure is equivalent to forming an $s \times n$ matrix with s successive blocks as rows and then transmit the symbols column by column. At the receiving end we put the received blocks into a matrix column by column and read the symbols out row by row.

By this procedure, adjacent symbols in any original block will be s symbols apart during transmission. Therefore, the memory of the channel can as a rule be decreased. In particular, using interleaving with an RFC channel model will again give an RFC channel model, but the new correlation coefficient is r^s, where r is the correlation coefficient of the original channel model.

Stochastic Outer Block Interleaving

Let π be some (pseudo-)randomly chosen permutation of ns elements. At the input, the ns symbols in s consecutive blocks are permuted using π. At the receiving end the inverse permutation π^{-1} is used.

Let $P(N, M)$ be the probability of having exactly M errors in a block of length N on the original channel. Then the probability $\tilde{p}(n, m)$ of an error pattern of length n and weight m after the transformation is given by

$$\tilde{p}(n, m) = \frac{\sum_{M=0}^{N} P(N, M) \binom{n}{m} \binom{N-n}{M-m}}{\binom{N}{M}}. \tag{1.38}$$

The transformed channel will be called a *quasi binomial channel* when the probability $P(i)$ of any error pattern $\bar{e}$ only depends on the error weight.

Inter Block Pseudorandom Interleaving

Inter block pseudorandom interleaving is the special case of stochastic outer block interleaving with $s = 1$, that is, the symbols within each block are permuted using some pseudo-randomly chosen permutation. The error multiplicity function is not changed after such a transformation.

The Stochastic Binary Additively Transformed Channel

For each new block $\bar{a} \in GF(2)^n$ to be transmitted, choose a random $\bar{\gamma} \in GF(2)^n$, compute

$$\bar{b} = \bar{a} + \bar{\gamma} \tag{1.39}$$

and transmit $\bar{b}$. At the receiving end, the received vector $\bar{\bar{b}}$ is added to $\bar{\gamma}$:

$$\bar{\bar{a}} = \bar{\bar{b}} + \bar{\gamma} \tag{1.40}$$

and

$$\bar{\bar{b}} = \bar{b} + \bar{e}, \tag{1.41}$$

where $\bar{e}$ is the channel error pattern that occurred during transmission.

We call this channel the *stochastic additively transformed channel*, SATC.

Let us show that if we use a random sequences $\bar{\gamma}$ which has equally likely and independent symbols, then any binary channel is transformed to a channel with additive noise. Let $P(\bar{\bar{b}}|\bar{b})$ denote the probability that $\bar{\bar{b}}$ is received when $\bar{b}$ is transmitted, and similarly $P(\bar{\bar{a}}|\bar{a})$ for the transformed $\bar{\bar{a}}$ at the receiving end given the untransformed $\bar{a}$ at the sending end. The assumptions on $\bar{\gamma}$ implies that

$$P(\bar{\gamma}) = 2^{-n}. \tag{1.42}$$

Further, $\bar{\gamma'} = \bar{a} + \bar{\gamma}$ runs through $GF(2)^n$ when $\bar{\gamma}$ does. Hence

$$\begin{aligned} P(\bar{\bar{a}}|\bar{a}) &= \sum_{\bar{\gamma} \in GF(2)^n} P(\bar{\gamma})P(\bar{a} + \bar{\gamma} + \bar{e}|\bar{a} + \bar{\gamma}) \\ &= 2^{-n} \sum_{\bar{\gamma'} \in GF(2)^n} P(\bar{\gamma'} + \bar{e}|\bar{\gamma'}). \end{aligned}$$

Therefore, $P(\bar{\bar{a}}|\bar{a})$ only depends on $\bar{e}$.

The Stochastic Multiplicatively Binary Transformed Channel

Let us divide the binary sequences $\bar{a}$ of symbols into blocks of the same lengths n. We consider the blocks as elements of the field $GF(2^n)$. Let $\bar{\gamma} \neq \bar{0}$ be a random sequence, again considered as an element of $GF(2^n)$. Let $\bar{a} \times \bar{\gamma}$ denote the product of these elements in $GF(2^n)$.

The *stochastic multiplicatively transformed channel*, SMTC, is defined similarly to the additively transformed channel, but the transformation used is

$$\begin{aligned} \bar{b} &= \bar{a} \times \bar{\gamma}, \\ \bar{\bar{a}} &= \bar{\bar{b}} \times \bar{\gamma}^{-1}, \end{aligned} \tag{1.43}$$

where, as above, $\bar{a}$, $\bar{b}$, $\bar{\bar{b}}$, and $\bar{\bar{a}}$ are the untransformed input block, the transformed input block, the received block, and the transformed received block, respectively.

Suppose that the original channel has additive noise. Let $P_{\text{ut}}(\bar{e})$ be the probability of the error pattern $\bar{e}$ in the (untransformed) channel and let $P_{\text{ef}}(n)$ be the probability of the all-zero error pattern on the (untransformed) channel. Then for the SMTC we get

$$P(\bar{e}) = \sum_{\bar{\gamma} \in GF(2^n)} P(\bar{\gamma})P_{\text{ut}}(\bar{\gamma} \times \bar{e}) = \begin{cases} P_{\text{ef}}(n) & \text{when } \bar{e} = \bar{0}, \\ (1 - P_{\text{ef}}(n))/(2^n - 1) & \text{when } \bar{e} \neq \bar{0}. \end{cases} \tag{1.44}$$

If we combine the two transforms SATC and SMTC we transform any channel (including one with non-additive noise) into a channel with additive noise given by (1.44). We see that this transformed channel has the same value for $P(\bar{e} = \bar{0})$ as the original channel, whereas all other error patterns are equally likely. We will use this property in Chapter 4 to design error detecting codes for channels with memory.

For an actual implementation, the $\bar{\gamma}$ will be pseudo-random sequences. In particular this require synchronization between the $\bar{\gamma}$ generated at the sending and the receiving end. A similar situation occurs with the application of stream ciphers in cryptography. It is more complicated to understand why (1.44) is true for transforms based on pseudo-random sequences. One can show that if the pseudo-random sequence it the output of a linear shift register of length n with a primitive polynomial as feedback, then (1.44) still is true. We will return to error detection with pseudo-random transforms in Chapter 4.

1.5 COMMENTS AND REFERENCES

Many of the quoted results are given in Korzhik and Fink [10].

The material on FPMC is from Bocharova and Kudryashov [41]. They considered the more general situation of Rice distribution of fadings.

(1.34)-(1.36) are from Turin [22].

More general results about properties of channels after pseudo-random transforms can be found in [61].

Generalizations of the BEC model have been described by Lu and Chang [141] and Wolf and Chun [207].

2

BASICS ON ERROR CONTROL

2.1 ABC ON LINEAR CODES

Basic Notations and Terminology

The basic idea of coding is to introduce redundancy that can be utilized to detect and, for some applications, correct errors that has occurred during a transmission over a channel. We will restrict our attention to encoding into vectors of symbols from a finite field of q elements, where q is some prime power; the most important case is $q = 2$. The reason for this restriction is that we can make use of the mathematical structure and extensive theory for vector spaces. However, it is quite possible to consider more general encodings.

Suppose that we have a set $\mathcal{M}$ of M possible messages that may be sent. An $(n, M; q)$ *code* is a subset of the vector space $GF(q)^n$ containing M vectors. An *encoding* is a one-to-one function from $\mathcal{M}$ to the code. The elements of the code are called *code words*.
On the symmetric channel, the most important codes are the linear codes. An $[n, k; q]$ *linear* code is a k-dimensional subspace of $GF(q)^n$. This is in particular an $(n, q^k; q)$ code.
When $q = 2$, we usually drop q from the notation and write just (n, M) and $[n, k]$ respectively.

Hamming Weight and Distance

The *Hamming weight* $w_{\mathrm{H}}(\bar{x})$ of a vector $\bar{x}$ over $GF(q)$ is the number of non-zero positions in $\bar{x}$, that is

$$w_{\mathrm{H}}(\bar{x}) = \#\{i \mid 1 \leq i \leq n \text{ and } x_i \neq 0\}.$$

The *Hamming distance* $d_{\mathrm{H}}(\bar{x}, \bar{y})$ between to vectors $\bar{x}, \bar{y} \in GF(q)^n$ is number of positions where they differ, that is

$$d_{\mathrm{H}}(\bar{x}, \bar{y}) = \#\{i \mid 1 \leq i \leq n \text{ and } x_i \neq y_i\}.$$

If a vector $\bar{x}$ is transmitted and e errors occurred during transmission, then the received vector $\bar{y}$ differs from $\bar{x}$ in e positions, that is $d_{\mathrm{H}}(\bar{x}, \bar{y}) = e$.

Clearly,

$$d_{\mathrm{H}}(\bar{x}, \bar{y}) = w_{\mathrm{H}}(\bar{x} - \bar{y}).$$

For an $(n, M; q)$ code C define the *minimum distance* by

$$d(C) = \min\{d_{\mathrm{H}}(\bar{x}, \bar{y}) \mid \bar{x}, \bar{y} \in C, \quad \bar{x} \neq \bar{y}\}.$$

Further

$$d(n, M; q) = \max\{d(C) \mid C \text{ is a } (n, M; q) \text{ code}\},$$

$$\delta(n, R; q) = \frac{d\left(n, \lceil q^{Rn} \rceil ; q\right)}{n},$$

and

$$\delta(R; q) = \limsup_{n \to \infty} \delta(n, R; q).$$

We note that R will be the rate of the codes considered. Sometimes we include $d = d(C)$ in the notation for a code and write $(n, M, d; q)$ and $[n, k, d; q]$.

Support of a Set of Vectors

For $\bar{x} = (x_1, x_2, \ldots, x_n) \in GF(q)^n$, we define its *support* by

$$\chi(\bar{x}) = \{i \in \{1, 2, \ldots, n\} \mid x_i \neq 0\}.$$

For a set $S \subseteq GF(q)^n$, we define its *support* by

$$\chi(S) = \bigcup_{\bar{x} \in S} \chi(\bar{x}).$$

Hence $\chi(S)$ is the set of positions where not all vectors in S are zero.

Extending Vectors

Let $\bar{x} = (x_1, x_2, \cdots, x_n) \in GF(q)^n$, $\bar{y} = (y_1, y_2, \cdots, y_m) \in GF(q)^m$ and $u \in GF(q)$. Then

$$u\bar{x} = (ux_1, ux_2, \cdots, ux_n),$$
$$(\bar{x}|u) = (x_1, x_2, \cdots, x_n, u),$$
$$(\bar{x}|\bar{y}) = (x_1, x_2, \cdots, x_n, y_1, y_2, \cdots, y_m).$$

For a subset S of $GF(q)^n$,

$$uS = \{u\bar{x} \mid \bar{x} \in S\},$$
$$(S|u) = \{(\bar{x}|u) \mid \bar{x} \in S\}.$$

For an $[n, k; q]$ code C and any r, where $1 \leq r \leq k$, *the r-th minimum support weight* is defined by

$$d_r = d_r(C) = \min\Big\{|\chi(D)| \ \Big| \ D \text{ is an } [n, r; q] \text{ subcode of } C\Big\}.$$

In particular, the minimum distance of C is d_1. The *weight hierarchy* of C is the set $\{d_1, d_2, \cdots, d_k\}$. The weight hierarchy satisfy the following inequality.

$$d_r \geq d_{r-1}\left(1 + \frac{q-1}{q^r - q}\right). \tag{2.1}$$

In particular, we have

$$d_r \geq d_{r-1} + 1. \tag{2.2}$$

Ordering

Let some ordering $\leq$ of $GF(q)$ be given. We extend this to a partial ordering of $GF(q)^n$ as follows:

$$(x_1, x_2, \cdots, x_n) \leq (y_1, y_2, \cdots, y_n) \text{ if } x_i \leq y_i \text{ for } 1 \leq i \leq n.$$

Entropy

The base b (or b-ary) *entropy function* $H_b(z)$ is defined by

$$H_b(z) = -z \log_b(z) - (1 - z) \log_b(1 - z)$$

for $0 \leq z \leq 1$. $H_b(z)$ is an increasing function on $\left[0, \frac{1}{2}\right]$, $H_b(0) = 0$, and $H_b\left(\frac{1}{2}\right) = \log_b(2)$. Define $\rho(z) = \rho_b(z)$ on $[0, \log_b(2)]$ by $\rho_b(z) \in \left[0, \frac{1}{2}\right]$ and

$$H_b(\rho_b(z)) = 1 - z.$$

Generator and Check Matrices

As already defined above, an $[n, k; q]$ *linear* code C is a k-dimensional subspace of $GF(q)^n$. Vector spaces can be represented in various ways. Suppose that $\{\bar{g}_1, \bar{g}_2, \cdots, \bar{g}_k\}$ is a basis for C. Then C is the set of all possible linear combinations of these vectors. Let G be the $k \times n$ matrix whose k rows are $\bar{g}_1, \bar{g}_2, \cdots, \bar{g}_k$. Then

$$C = \{\bar{x}G \mid \bar{x} \in GF(q)^k\}.$$

We call G a *generator matrix* for C. A natural *encoding* $GF(q)^k \to GF(q)^n$ is given by

$$\bar{x} \mapsto \bar{x}G.$$

If $T : GF(q)^k \to GF(q)^k$ is a linear invertible transformation, then TG is also a generator matrix, this is just a change of basis.

If H is a generator matrix for the orthogonal space

$$C^{\perp} = \{\bar{x} \in GF(q)^n \mid \bar{x}\bar{c}^t = 0 \text{ for all } \bar{c} \in C\}$$

(this is known as the *dual code*), then

$$C = \{\bar{x} \in GF(q)^n \mid \bar{x}H^t = 0\},$$

where H^t is the transposed of H. H is known as a *(parity) check matrix* for C. Note that $GH^t = 0$ and that any $(n-k) \times n$ matrix H of rank $n-k$ such that $GH^t = 0$ is a check matrix.

Systematic Linear Codes

A code C is called *systematic* if it has a generator matrix G such that

$$\bar{x}G = (\bar{x}|\bar{z}(\bar{x}))$$

for all $\bar{x} \in GF(q)^k$, that is $G = (I_k|P)$ where I_k is the identity matrix and P is some $k \times (n-k)$ matrix. The point is that the information $\bar{x}$ to be encoded appears as the first part of the encoded vector $\bar{x}G$.

We have used standard terminology, however, we point out that it would be more appropriate to say that the encoding is systematic; a generator matrix $G' \neq G$ for the same code does not have the property that the information $\bar{x}$ appears as the first part of the encoded vector $\bar{x}G'$.

Since
$$(I_k|P)(-P^t|I_{n-k})^t = -P + P = 0,$$
$H = (-P^t|I_{n-k})$ is a check matrix for C.

Equivalent Codes

Two $[n, k]$ codes C_1, C_2 are *equivalent* if

$$C_2 = \{\bar{x}Q\Pi \mid \bar{x} \in C_1\}$$

where Q is a non-singular diagonal $n \times n$ matrix and Π is an $n \times n$ permutation matrix. If G is a generator matrix for an C_1 code, then $GQ\Pi$ is a generator matrix for C_2. We note that equivalent codes have the same weight distribution, and in particular the same minimum distance.

Let G be a $k \times n$ generator matrix. By suitable row operations this can be brought into reduced echelon form. A suitable permutation of the columns will give a matrix of the form $(I_k|P)$. In other words there exists an invertible $k \times k$ matrix T and an $n \times n$ permutation matrix Π such that

$$TG\Pi = (I_k|P).$$

Therefore, any code is equivalent to a systematic code.

New Codes From Old

There are a number of ways to construct new codes from one or more old ones. We will describe some of these briefly. In a later section we will discuss how the error detecting capability of the new codes are related to the error detecting capability of the old ones.

*The *-operation*

Let S_k be a $k \times \frac{q^k-1}{q-1}$ matrix over $GF(q)$ such that
 (i) all columns of S_k are non-zero,
 (ii) if $\bar{x} \neq \bar{y}$ are columns, then $\bar{x} \neq j\bar{y}$ for all $j \in GF(q)$.

Let C be an $[n, k; q]$ code with generator matrix G. The $[n + \frac{q^k-1}{q-1}, k]$ code with generator matrix $(G|S_k)$ will be denoted by C^*. We remark that the matrix S_k is in no way unique, and two different choices for S_k will in general give two different codes under this construction. These codes are equivalent however.

We also consider iterations of the *-operation. We define C^{r*} by

$$\begin{aligned} C^{0*} &= C, \\ C^{(r+1)*} &= (C^{r*})^* . \end{aligned}$$

Extending a Code

Consider an $[n, k; q]$ code C. Let $\bar{b} = (b_1, b_2, \cdots, b_n) \in GF(q)^n$. Let C^{ex} be the $[n+1, k; q]$ code

$$C^{\text{ex}} = \left\{ (a_1, a_2, \cdots, a_n, -\sum_{i=1}^{n} a_i b_i) \,\middle|\, (a_1, a_2, \cdots, a_n) \in C \right\}.$$

Let H be a check matrix for C. A check matrix for C^{ex} is

$$\begin{pmatrix} H & \bar{0}^t \\ \bar{b} & 1 \end{pmatrix}.$$

In particular, in the binary case, if $b_1 = b_2 = \cdots = b_n = 1$, we have extended the code with a parity check. The code $(GF(2)^n)^{\text{ex}}$ is known as the ***single parity check code*** or just the parity check code.

Puncturing a Code

Puncturing is to remove a particular position from each code word. The dimension of the code may or may not decrease by 1. Clearly, the operation of puncturing may be repeated.

Zero-sum Subcodes of a Code

Consider an $[n, k; q]$ code C. The *zero-sum subcode* C^{zs} is the code

$$C^{zs} = \left\{(a_1, a_2, \cdots, a_n) \in C \;\middle|\; \sum_{i=1}^{n} a_i = 0\right\}.$$

If C contains code words such that $\sum_{i=1}^{n} a_i \neq 0$, then the zero-sum subcode is an $[n, k-1; q]$ code.

In the binary case $\sum_{i=1}^{n} a_i = 0$ if and only if $w_H(\bar{a})$ is even, and C^{zs} is then called the *even-weight subcode.*

Shortening a Code

Consider an $[n, k; q]$ code C with the first position in its support. *Shortening* (by the first position) we obtain the $[n-1, k-1]$ code

$$C^{s} = \left\{\bar{x} \in GF(2)^{n-1} \;\middle|\; (0|\bar{x}) \in C\right\},$$

that is, we take the set of all code words of C with 0 in the first position and remove that position. More general, we can shorten by any position.

If $G = (I_k|P)$ generates a systematic code, then a generator matrix for the shortened code is obtained by removing the first row and the first column of G.

We note that shortening will not decrease the minimum distance; however it may increase.

Product Codes

Let C_1 be an $[n_1, k_1, d_1; q]$ code and C_2 an $[n_2, k_2, d_2; q]$ code. The *product code* is the $[n_1 n_2, k_1 k_2, d_1 d_2; q]$ code C whose code words are usually written as an $n_1 \times n_2$ array; C is the set of all such arrays where all rows belongs to C_1 and all columns to C_2.

Concatenated Codes

Codes can be concatenated in various ways. One such construction that has been proposed for a combined error correction and detection is the following.

Let C_1 be an $[N, K; q]$ code and C_2 an $[n, k; q]$ code, where $N = mk$ for some integer m. The encoding is done as follows: K information symbols is encoded into N symbols using code C_1. These $N = mk$ are split into m blocks with k symbols in each block. Then each block is encoded into n symbols using code C_2. The concatenated code is an $[mn, K; q]$ code. If G_1 and G_2 are generator matrices for C_1 and C_2 respectively, then a generator matrix for the combined code is the following.

$$G_1 \begin{pmatrix} G_2 & 0 & \cdots & 0 \\ 0 & G_2 & \cdots & 0 \\ \vdots & \vdots & \ddots & \vdots \\ 0 & 0 & \cdots & G_2 \end{pmatrix}.$$

The construction above can be generalized in various ways. One generalization that is used in several practical systems combines a convolutional code for error correction and a block code (e.g. an CRC code) for error detection.

Cyclic and Shortened Cyclic Codes

A code $C \subseteq GF(q)^n$ is called *cyclic* if

$$(a_{n-1}, a_{n-2}, \cdots, a_0) \in C \text{ implies that } (a_{n-2}, a_{n-3}, \cdots, a_0, a_{n-1}) \in C.$$

Our reason for the special way of indexing the elements is that we want to associate a polynomial in the variable z with each n-tuple as follows:

$$\bar{a} = (a_{n-1}, a_{n-2}, \cdots, a_0) \leftrightarrow \bar{a}(z) = \sum_{i=0}^{n-1} a_i z^i.$$

This correspondence has the following property (it is an isomorphism): if $\bar{a}, \bar{b} \in GF(q)^n$ and $c \in GF(q)$, then

$$\begin{aligned} \bar{a} + \bar{b} &\leftrightarrow \bar{a}(z) + \bar{b}(z), \\ c\bar{a} &\leftrightarrow c\bar{a}(z). \end{aligned}$$

In particular, any code may be represented as a set of polynomials. Moreover, the polynomial corresponding to $(a_{n-2}, a_{n-3}, \cdots, a_0, a_{n-1})$ is

$$a_{n-1} + \sum_{i=0}^{n-2} a_i z^{i+1} = z\bar{a}(z) - a_{n-1}(z^n - 1) = [z\bar{a}(z) \pmod{z^n - 1}].$$

In the rest of this section we use this code representation.

Many important codes are cyclic codes or shortened cyclic codes. One reason that cyclic codes are used is that they have more algebraic structure then linear codes in general, and this structure can be used both in the analysis of the codes and in the design of efficient encoders and decoders for error correction. For example, the roots of the polynomial $g(z)$ given by the theorem below give information on the minimum distance of the code. Hamming codes is one class of cyclic codes and shortened Hamming codes and their cosets are used in several standards for data transmission where error detection is important. This is our main reason for introducing them in this text.

Theorem 2.1.1 *Let C be a cyclic $[n, k; q]$ code. Then there exists a monic polynomial $g(z)$ of degree $n - k$ such that*

$$C = \{v(z)g(z) \mid \deg(v(z)) < k\}.$$

Proof: Let $g(z)$ be the monic polynomial in C of smallest positive degree, say degree m. Then $z^i g(z) \in C$ for $0 \leq i < n - m$. Let $a(z)$ be any non-zero polynomial in C, of degree s say; $n > s \geq m$. Then there exist elements $c_{s-m}, c_{s-m-1}, \cdots, c_0 \in GF(q)$ such that

$$r(z) = a(z) - \sum_{i=0}^{s-m} c_i z^i g(z)$$

has degree less than m (this can easily be shown by induction on s). Since C is a linear code, $r(z) \in C$. Moreover, there exists a $c \in GF(q)$ such that $cr(z)$ is monic, and the minimality of the degree of $g(z)$ implies that $r(z) \equiv 0$. Hence $a(z) = v(z)g(z)$ where $v(z) = \sum_{i=0}^{s-m} c_i z^i$. In particular, the set

$$\{g(z), zg(z), \cdots, z^{n-1-m}g(z)\}$$

of $n - m$ polynomials is a basis for C and so $n - m = k$, that is $r = n - m$.

The polynomial $g(z)$ is called the *generator polynomial* of C.

If $g(1) \neq 0$, then the code generated by $(z-1)g(z)$ is an $[n+1, k; q]$ code. It is the code C^{ex} obtained from C extending using the vector $\bar{1} = (11 \cdots 1)$, that is

$$\Big\{(a_1, a_2, \cdots, a_n, -\sum_{i=1}^{n} a_i) \Bigm| (a_1, a_2, \cdots, a_n) \in C\Big\}.$$

Encoding using a cyclic code is usually done in one of two ways. Let $\bar{v} = (v_{k-1}, v_{k-2}, \cdots, v_0) \in GF(q)^k$ be the information to be encoded. The first, and direct way of encoding, is to encode into $\bar{v}(z)g(z)$. On the other hand, the code is systematic, but this encoding is not. The other way of encoding is to encode $\bar{v}(z)$ into the polynomial in C "closest" to $z^{n-k}\bar{v}(z)$. More precisely, there is a unique $\bar{a}(z)$ of degree less then k such that

$$\bar{r}(z) = z^{n-k}\bar{v}(z) - \bar{a}(z)g(z)$$

has degree less then $n-k$, and we encode into

$$\bar{a}(z)g(z) = z^{n-k}\bar{v}(z) + \bar{r}(z).$$

The corresponding code word has the form $(\bar{v}|\bar{r})$, where $\bar{r} \in GF(q)^{n-k}$.

Theorem 2.1.2 *Let C be a cyclic $[n, k; q]$ code with generator polynomial $g(z)$. Then $g(z)$ divides $z^n - 1$, that is, there exists a monic polynomial $h(z)$ of degree k such that*

$$g(z)h(z) = z^n - 1,$$

Moreover, the polynomial

$$\tilde{h}(z) = -g(0)z^k h\left(\frac{1}{z}\right)$$

is the generator polynomial of $C^{\perp}$.

Proof: There exist unique polynomials $h(z)$ and $r(z)$ such that

$$z^n - 1 = g(z)h(z) + r(z)$$

and $\deg(r(z)) < n - k$. In particular $r(z) \equiv h(z)g(z) \pmod{z^n - 1}$ and so $r(z) \in C$. The minimality of the degree of $g(z)$ implies that $r(z) \equiv 0$.

Let $g(z) = \sum_{i=0}^{n-k} g_i z^i$ and $h(z) = \sum_{i=0}^{k} h_i z^i$. Then

$$\sum_i g_{l-i}h_i = \begin{cases} -1 & \text{if } l = 0, \\ 0 & \text{if } 0 < l < n, \\ 1 & \text{if } l = n. \end{cases}$$

Further, $\tilde{h}(z) = -g_0 \sum_{i=0}^{k} h_{k-i} z^i$. Since $-g_0 h_0 = 1$, $\tilde{h}(z)$ is monic. Let

$$\bar{v} = (0, 0, \cdots, 0, g_{n-k}, \cdots, g_0),$$

$$\bar{u} = (0, 0, \cdots, 0, h_0, \cdots, h_k),$$

and $\bar{v}^l, \bar{u}^l$ the vectors l times cyclicly shifted, that is

$$\bar{u}^l = (h_{k-l+1}, h_{k-l+2}, \cdots, h_k, 0, \cdots, 0, h_0, h_1, \cdots, h_{k-l}),$$

and $\bar{v}^l$ similarly. First, we see that

$$\bar{v} \cdot \bar{u}^l = \sum_{i=0}^{k} g_{k+l-i} h_i = 0$$

for $-k < l < n - k$. Hence,

$$\bar{v}^m \cdot \bar{u}^l = \bar{v} \cdot \bar{u}^{l-m} = 0$$

for $0 \leq m < k$ and $0 \leq l < n - k$; that is, each basis vector for C is ortogonal to each basis vector in the code $\tilde{C}$ generated by $\tilde{h}(z)$, and so $\tilde{C} = C^{\perp}$.

The polynomial $g(z)$ of degree m is called *primitive* if the least positive n such that $g(z)$ divides $z^n - 1$ is $q^m - 1$. The cyclic code C generated by a primitive $g(z)$ of degree m is a $\left[\frac{q^m-1}{q-1}, \frac{q^m-1}{q-1} - m; q\right]$ Hamming code.

The code obtained by shortening the cyclic $[n, k; q]$ code C m times is the $[n - m, k - m; q]$ code

$$\{\bar{v}(z)g(z) \mid \deg(\bar{v}(z)) < k'\} \tag{2.3}$$

where $k' = k - m$. Note that (2.3) defines an $[n - k + k', k'; q]$ code for all $k' > 0$, not only for $k' \leq k$. These codes are also known as *cyclic redundancy-check* (CRC) codes. The dual of an $[n - k + k', k'; q]$ code C generated by $g(z) = \sum_{i=0}^{n-k} g_i z^i$ where $g_{n-k} = 1$ can be described as a systematic code as follows: The information sequence $(a_{n-k-1}, \ldots, a_0)$ is encoded into the sequence $(a_{n-k-1+k'}, \ldots, a_0)$ where

$$a_j = - \sum_{i=0}^{n-k-1} g_i a_{j-n+k+i}.$$

This follows from the fact that

$$(a_{n-k-1+k'}, a_{n-k-2+k'}, \ldots, a_0) \cdot (0, 0, \ldots, 0, g_{n-k}, \ldots, g_0, \overbrace{0, \ldots, 0}^{i}) = 0$$

for $0 \le i \le k'$ by definition and that $(0, 0, \ldots, 0, g_{n-k}, \ldots, g_0, \overbrace{0, \ldots, 0}^{i})$ where $0 \le i \le k'$ is a basis for C.

A number of binary CRC codes have become international standards for error detection in various contexts; in particular CRC codes generated by the following polynomials of degree 16:

IEC TC57	$z^{16} + z^{14} + z^{12} + z^{11} + z^9 + z^8 + z^7 + z^4 + z + 1$,
IEEE WG77.1	$z^{16} + z^{14} + z^{13} + z^{11} + z^{10} + z^9 + z^8 + z^6 + z^5 + z + 1$,
CCITT X.25	$z^{16} + z^{12} + z^5 + 1$,
ANSI	$z^{16} + z^{15} + z^2 + 1$,
IBM-SDLC	$z^{16} + z^{15} + z^{13} + z^7 + z^4 + z^2 + z + 1$.

We will return to a more detailed discussion of these and other CRC codes in the section 3.10.

2.2 WEIGHT DISTRIBUTIONS

Distance and Weight Distribution

Let C be an $(n, M; q)$ code. Let

$$A_i = A_i(C) = \frac{1}{M} \#\{(\bar{x}, \bar{y}) \mid \bar{x}, \bar{y} \in C \text{ and } d_{\mathrm{H}}(\bar{x}, \bar{y}) = i\}.$$

The sequence $A_0, A_1, \cdots, A_n$ is known as the *distance distribution* of C and

$$A_C(z) = \sum_{i=0}^{n} A_i z^i$$

is the *distance distribution function* of C.

Let

$$A_i^{\mathrm{w}} = A_i^{\mathrm{w}}(C) = \#\{\bar{x} \in C \mid w_{\mathrm{H}}(\bar{x}) = i\}.$$

The sequence $A_0^{\mathrm{w}}, A_1^{\mathrm{w}}, \cdots, A_n^{\mathrm{w}}$ is known as the *weight distribution* of C and

$$A_C^{\mathrm{w}}(z) = \sum_{i=0}^{n} A_i^{\mathrm{w}} z^i$$

is the *weight distribution function* of C.

We note that $d_H(\bar{x},\bar{y}) = w_H(\bar{x}-\bar{y})$. If C is linear, then $\bar{x}-\bar{y} \in C$ when $\bar{x},\bar{y} \in C$. Hence for linear codes we get $A_i = A_i^w$, for all i and so $A_C(z) = A_C^w(z)$. For convenience, we drop the w from the notation for linear codes and call $A_C(z)$ the weight distribution.

If C and C' are equivalent linear codes, then clearly $A_i(C) = A_i(C')$. For the study of the weight distribution of linear codes we may therefore without loss of generality assume that the code is systematic if we so wish.

MacWilliams Transform

Let C be an $(n, M; q)$ code. The *MacWilliams transform* of $A_C(z)$ is defined by

$$A_C^{\mathrm{MW}}(z) = \frac{1}{M}(1+(q-1)z)^n A_C\left(\frac{1-z}{1+(q-1)z}\right).$$

We denote the coefficients of $A_C^{\mathrm{MW}}(z)$ by $B_i = B_i(C)$, that is,

$$A_C^{\mathrm{MW}}(z) = \sum_{i=0}^{n} B_i z^i.$$

Substituting $\frac{1-z}{1+(q-1)z}$ for z we get the following lemma.

Lemma 2.2.1

$$A_C(z) = \frac{M}{q^n}(1+(q-1)z)^n A_C^{\mathrm{MW}}\left(\frac{1-z}{1+(q-1)z}\right).$$

Two important relations are the following.

Theorem 2.2.1 *Let C be an $(n, M; q)$ code. Then*

$$B_i(C) \geq 0 \text{ for } 0 \leq i \leq n.$$

In particular, $B_0(C) = 1$.

Theorem 2.2.2 *Let C be a linear $[n, k; q]$ code. Then*

$$A_C^{\mathrm{MW}}(z) = A_{C^\perp}(z).$$

Differentiating the definition of $A_C^{\mathrm{MW}}(z)$ s times and putting $z=1$ we get the following relations.

Theorem 2.2.3 *Let C be an $(n, M; q)$ code and $s \geq 0$. Then*

$$\sum_{i=0}^{n} B_i \binom{i}{s} = \frac{q^{n-s}}{M} \sum_{j=0}^{s} A_j (-1)^j (q-1)^{s-j} \binom{n-j}{s-j}.$$

In particular, if C is an $[n, k, d; q]$ code and $s < d$, then

$$\sum_{i=0}^{n} B_i \binom{i}{s} = q^{n-k-s} (q-1)^s \binom{n}{s}.$$

A Generalized Weight Distribution

There are many generalizations of the weight distribution. One that is particularly important for error detection is the following.

Let C be an $[n, k]$ code and m a divisor of n. Let $A_{i_1, i_2, \cdots, i_m}(C)$ be the number of vectors $(\bar{x}_1 | \bar{x}_2 | \cdots | \bar{x}_m) \in C$ such that each part $\bar{x}_j \in GF(q)^{n/m}$ and $w_{\mathrm{H}}(\bar{x}_j) = i_j$ for $j = 1, 2, \cdots, m$. Further, let

$$A_C(z_1, z_2, \cdots, z_m) = \sum_{i_1, i_2, \cdots, i_m} A_{i_1, i_2, \cdots, i_m}(C) z_1^{i_1} z_2^{i_2} \cdots z_m^{i_m}.$$

For $m = 1$ we get the usual weight distribution function. Theorem 2.2.2 generalizes as follows.

Theorem 2.2.4 *Let C be a linear $[n, k; q]$ code. Then*

$$A_C(z_1, z_2, \cdots, z_m) = q^{k-n} \left\{ \prod_{j=1}^{m} (1 + (q-1) z_j) \right\}^{\frac{n}{m}} A_{C^\perp}(z_1', z_2', \cdots z_m')$$

where

$$z_j' = \frac{1 - z_j}{1 + (q-1) z_j}.$$

Codes Over Larger Fields

There is an alternative expression for the weight distribution function that is useful for some applications. Let G be a $k \times n$ generator matrix over $GF(q)$. Let $m_G : GF(q)^k \to \mathcal{N} = \{0, 1, 2, \ldots\}$, the *column count function*, be defined such that G contains exactly $m(\bar{x}) = m_G(\bar{x})$ columns equal to $\bar{x}$ for all $\bar{x} \in GF(q)^k$. We use the following further notations:

$[a]_b = \prod_{i=0}^{b-1}(q^a - q^i)$,

$s(U, m) = \sum_{\bar{x} \in U} m(\bar{x})$ for all $U \subseteq GF(q)^k$,

$\mathcal{S}_{kl}$ is the set of l dimensional subspaces of $GF(q)^k$,

$\sigma_{kl}(m, z) = \sum_{U \in \mathcal{S}_{kl}} z^{s(U^c, m)}$, where $U^c = GF(q)^k \setminus U$, the compliment of U,

$\hat{U} = \{\bar{y} \in GF(q^r)^k \mid \bar{y} \cdot \bar{x} = 0 \text{ for } \bar{x} \in GF(q)^k \text{ if and only if } \bar{x} \in U\}$,

$C_r = \{\bar{y}G \mid \bar{y} \in GF(q^r)^k\}$, the code generated by G over $GF(q^r)$,

$C = C_1$.

Theorem 2.2.5 *For $r \geq 1$ we have*

$$A_{C_r}(z) = \sum_{l=0}^{k} [r]_{k-l} \sigma_{kl}(m, z).$$

Proof: First we note that if $\bar{y} \in \hat{U}$, then

$$w_{\mathrm{H}}(\bar{y}G) = \sum_{\bar{x} \in GF(q)^k} m(\bar{x}) w_{\mathrm{H}}(\bar{y} \cdot \bar{x}) = \sum_{\bar{x} \in U^c} m(\bar{x}) = s(U^c, m).$$

Hence

$$A_{C_r}(z) = \sum_{l=0}^{k} \sum_{U \in \mathcal{S}_{kl}} \sum_{\bar{y} \in \hat{U}} z^{w_{\mathrm{H}}(\bar{y}G)} = \sum_{l=0}^{k} \sum_{U \in \mathcal{S}_{kl}} z^{s(U^c, m)} \sum_{\bar{y} \in \hat{U}} 1.$$

Since

$$\sum_{\bar{y} \in \hat{U}} 1 = [r]_{k-l},$$

the theorem follows.

For $r = 1$, we get the following alternative expression for the weight distribution of C.

Corollary 2.2.1 *We have*

$$A_C(z) = 1 + \sum_{U \in \mathcal{S}_{k,k-1}} z^{s(U^c,m)}.$$

Weight Distribution of Cosets

Next we give some results on the weight distribution of a coset of a code.

Theorem 2.2.6 *Let C be an $[n,k;q]$ code and S a proper coset of C. Let D be the $[n,k+1;q]$ code containing C and S. Then*

$$A_S^{\mathrm{w}}(z) = \frac{1}{q-1}\Big\{A_D(z) - A_C(z)\Big\}.$$

Proof: For each non-zero $a \in GF(q)$, $aS = \{a\bar{x} \mid \bar{x} \in S\}$ is also a proper coset of C and $A_{aS}^{\mathrm{w}}(z) = A_S^{\mathrm{w}}$. Further $D = C \cup \bigcup_{a \neq 0} aS$ (disjoint union) and so

$$A_D(z) = A_C(z) + (q-1)A_S^{\mathrm{w}}(z), \tag{2.4}$$

and the theorem follows.

Using the MacWilliams identity we get the following alternative expression.

Corollary 2.2.2 *Let C be an $[n,k;q]$ code and S a proper coset of C. Let D be the $[n,k+1;q]$ code containing C and S. Then*

$$A_S^{\mathrm{w}}(z) = \frac{\Big(1+(q-1)z\Big)^n}{q^{n-k}(q-1)}\left\{qA_D^{\perp}\left(\frac{1-z}{1+(q-1)z}\right) - A_C^{\perp}\left(\frac{1-z}{1+(q-1)z}\right)\right\}.$$

Theorem 2.2.7 *Let C be an $[n,k;q]$ code and S a proper coset of C. Then*

$$A_S^{\mathrm{w}}(z) \geq z^{n-k}A_C(z) \tag{2.5}$$

for all $z \in [0,1]$.

Proof: We may assume without loss of generality that the code C is systematic. There exists a $\bar{v} \in S$ such that $S = \bar{v} + C$ and such that $\bar{v} = (\bar{0}|\bar{b})$ where $\bar{b} \in GF(q)^{n-k}$.

Let $(\bar{x}|\bar{x}') \in C$ where $\bar{x} \in GF(q)^k$ and $\bar{x}' \in GF(q)^{n-k}$. Then

$$\begin{aligned} w_H((\bar{x}|\bar{x}') + (\bar{0}|\bar{b})) &= w_H(\bar{x}) + w_H(\bar{x}' + \bar{b}) \\ &\le w_H(\bar{x}) + n - k \\ &\le w_H((\bar{x}|\bar{x}')) + n - k \end{aligned}$$

and so

$$z^{w_H((\bar{x}|\bar{x}')+(\bar{0}|\bar{b}))} \ge z^{n-k} z^{w_H((\bar{x}|\bar{x}'))}.$$

Summing over all $(\bar{x}|\bar{x}') \in C$, the theorem follows.

Corollary 2.2.3 *Let C be an $[n, k; q]$ code and D an $[n, k+1; q]$ code containing C. Then*

$$A_D(z) \ge \left\{1 + (q-1)z^{n-k}\right\} A_C(z).$$

Proof: Let $S \subset D$ be a proper coset of C. By (2.4) and Theorem 2.2.7 we have

$$A_D(z) = A_C(z) + (q-1)A_S^{w}(z) \ge A_C(z) + (q-1)z^{n-k}A_C(z).$$

Theorem 2.2.8 *Let C be an $[n, k; q]$ code and S a proper coset of C. Then*

$$A_S^{w}(z) \le \frac{1 - y^{k+1}}{1 + (q-1)y^{k+1}} A_C(z)$$

for all $z \in [0, 1]$, where $y = (1-z)/(1+(q-1)z)$.

Theorem 2.2.9 *Let C be an $[n, k; q]$ code and D an $[n, k+1; q]$ which contains C. Then*

$$A_C(z) \ge \frac{1 + (q-1)y^{k+1}}{q} A_D(z)$$

for all $z \in [0, 1]$, where $y = (1-z)/(1+(q-1)z)$.

Proof: By Corollary 2.2.3 we get

$$\begin{aligned} A_C(z) &= q^{k-n}\Big(1+(q-1)z\Big)^n A_{C^\perp}(y) \\ &\geq q^{k-n}\Big(1+(q-1)z\Big)^n \Big(1+(q-1)y^{k+1}\Big) A_{D^\perp}(y) \\ &= q^{-1}\Big(1+(q-1)y^{k+1}\Big) A_D(z) \\ &= q^{-1}\Big(1+(q-1)y^{k+1}\Big)\Big(A_C(z)+(q-1)A_S^{\mathrm{w}}(z)\Big) \end{aligned}$$

and the theorems follow.

Corollary 2.2.4 *If C is an $[n,k;q]$ code and $k<n$, then*

$$A_C(z) \geq \frac{(1+(q-1)z)^n}{q^{n-k}} \prod_{j=k+1}^{n} \left(1+(q-1)y^j\right),$$

for all $z \in [0,1]$, where $y = (1-z)/(1+(q-1)z)$.

Proof: The corollary follows from Theorem 2.2.9 by induction on k.

Counting Vectors in a Sphere

The *sphere* $S_t(\bar{x})$ of radius t around a vector $\bar{x} \in GF(q)^n$ is the set of vectors within Hamming distance t of $\bar{x}$, that is

$$S_t(\bar{x}) = \{\bar{y} \in GF(q)^n \mid d_{\mathrm{H}}(\bar{x},\bar{y}) \leq t\}.$$

Let $N_t(i,j)$ be the number of vectors of weight j in a sphere of radius t around a vector of weight i.

Theorem 2.2.10 *We have*

$$N_t(i,j) = \sum_{e=|i-j|}^{t} \sum_{\delta=\max(i,j)-e}^{\min(\lfloor\frac{i+j-e}{2}\rfloor, n-e)} \binom{n-i}{\beta} \frac{i!}{\gamma!\delta!\epsilon!}(q-1)^\beta (q-2)^\epsilon$$

where $\beta = e-i+\delta$, $\gamma = e-j+\delta$, $\epsilon = i+j-e-2\delta$.

Proof: Let $w_{\rm H}(\bar{x}) = i$ and let $\bar{y} \in S_t(\bar{x})$ such that $w_{\rm H}(\bar{y}) = j$. Let

$$\begin{aligned}
\alpha &= \#\{l \mid x_l = y_l = 0\},\\
\beta &= \#\{l \mid x_l = 0, y_l \neq 0\},\\
\gamma &= \#\{l \mid x_l \neq 0, y_l = 0\},\\
\delta &= \#\{l \mid x_l = y_l \neq 0\},\\
\epsilon &= \#\{l \mid x_l \neq 0, y_l \neq 0, x_l \neq y_l\}.
\end{aligned} \tag{2.6}$$

Then

$$\begin{aligned}
i &= w_{\rm H}(\bar{x}) = \gamma + \delta + \epsilon,\\
j &= w_{\rm H}(\bar{y}) = \beta + \delta + \epsilon,\\
e &= d_{\rm H}(\bar{x}, \bar{y}) = \beta + \gamma + \epsilon,\\
n &= \alpha + \beta + \gamma + \delta + \epsilon.
\end{aligned} \tag{2.7}$$

Hence

$$\begin{aligned}
\beta &= e - i + \delta,\\
\gamma &= e - j + \delta,\\
\epsilon &= i + j - e - 2\delta.
\end{aligned} \tag{2.8}$$

Further,

$$\begin{aligned}
&|i - j| \leq e \leq t,\\
&\delta = i - e + \beta \geq i - e,\\
&\delta = j - e + \gamma \geq j - e,\\
&\delta = n - e - \alpha \leq n - e,\\
&2\delta = i + j - e - \epsilon \leq i + j - e.
\end{aligned} \tag{2.9}$$

On the other hand, if e and δ are integers such that (2.9) is satisfied, then there are

$$\binom{n-i}{\beta} \frac{i!}{\gamma!\delta!\epsilon!} (q-1)^{\beta} (q-2)^{\epsilon}$$

ways to choose $\bar{y}$ such that (2.6)-(2.8) is satisfied.

For $q = 2$, the terms in the sum for $N_t(i,j)$ are 0 unless $\epsilon = 0$. We get the following simpler expression in this case:

$$N_t(i,j) = \sum_{\gamma=\max(0,i-j)}^{\lfloor \frac{t+i-j}{2} \rfloor} \binom{n-i}{\gamma+j-i} \binom{i}{\gamma}. \tag{2.10}$$

Bounds on the Number of Code Words of a Given Weight

Some useful upper bounds on A_i for a linear code are given by the next theorem.

Theorem 2.2.11 *Let C be a linear $[n,k,d=2t+1;q]$ code. If $N_t(i,j)>0$, then*

$$A_i \leq \frac{\binom{n}{j}}{N_t(i,j)}(q-1)^j.$$

In particular, for $d \leq i \leq \lfloor \frac{n}{2} \rfloor$ we have

$$A_i \leq \frac{\binom{n}{i}}{\binom{n-i+t}{t}}(q-1)^{i-t} \leq \frac{\binom{n}{i}}{\binom{\lceil \frac{n}{2} \rceil + t}{t}}(q-1)^{i-t},$$

and, for $\lceil \frac{n}{2} \rceil \leq i \leq n-t$,

$$A_i \leq \frac{\binom{n}{i}}{\binom{i+t}{t}}(q-1)^i \leq \frac{\binom{n}{i}}{\binom{\lceil \frac{n}{2} \rceil + t}{t}}(q-1)^i.$$

Proof: Counting all vectors of weight j and Hamming distance at most t from a code word of weight i we get

$$A_i N_t(i,j) \leq \binom{n}{j}(q-1)^j.$$

In particular, $N_t(i,i-t) = \binom{i}{t} > 0$ for all $i \geq d$, and so

$$A_i \leq \frac{\binom{n}{i-t}}{\binom{i}{t}}(q-1)^{i-t} = \frac{\binom{n}{i}}{\binom{n-i+t}{t}}(q-1)^{i-t}.$$

Similarly, $N_t(i,i+t) = \binom{n-i}{t}(q-1)^t > 0$ for $d \leq i \leq n-t$ and so

$$A_i \leq \frac{\binom{n}{i+t}(q-1)^{i+t}}{\binom{n-i}{t}(q-1)^t} = \frac{\binom{n}{i}}{\binom{i+t}{t}}(q-1)^i.$$

Theorem 2.2.12 *For an $[n,k;q]$ code C we have $A_n \leq (q-1)^k$.*

Proof: Since equivalent codes have the same weight distribution, we may assume without loss of generality that the code is systematic, that is, it is generated by a matrix

$$G = (I_k|P) = \begin{pmatrix} \bar{g}_1 \\ \vdots \\ \bar{g}_k \end{pmatrix}$$

where I_k is the $k \times k$ identity matrix, P is a $k \times (n-k)$ matrix, and $\bar{g}_1, \ldots, \bar{g}_k$ are the rows of G. If $\bar{c} = \sum_{i=1}^{k} a_i \bar{g}_i$ has weight n, then in particular $a_i = c_i \neq 0$ for $1 \leq i \leq k$. Hence there are at most $(q-1)^k$ such $\bar{c}$.

Extending a Code

The matrix S_k and the *-operation were defined on page 22. The code generated by S_k is a constant weight code, that is, all non-zero code words have the same weight, namely q^{k-1}.

Therefore, $A_{C^*}(z)$ only depends on $A_C(z)$. In fact

$$A_{C^*}(z) - 1 = z^{q^{k-1}}(A_C(z) - 1)$$

since each non-zero vector is extended by a part of weight q^{k-1}.

2.3 PRINCIPLES OF ERROR DETECTION

Consider what happens when a code word $\bar{x}$ from an (n, M) code C is transmitted over a channel K and errors occurs during transmission. If the received vector $\bar{y}$ is not a code word we immediately realize that something has gone wrong during transmission, we *detect* that errors have occurred. However, it may happen that the combination of errors is such that the received vector $\bar{y}$ is also a code word. In this case we have no way to tell that the received code word is not the sent code word. Therefore, we have an *undetectable error*. We let $P_{\text{ue}} = P_{\text{ue}}(C, K)$ denote the probability that this happens. If $P(\bar{x})$ is the probability that $\bar{x}$ was sent and $P(\bar{y}|\bar{x})$ is the probability that $\bar{y}$ is received, given that $\bar{x}$ was sent, then

$$P_{\text{ue}}(C, K) = \sum_{\bar{x} \in C} P(\bar{x}) \sum_{\bar{y} \in C \setminus \{\bar{x}\}} P(\bar{y}|\bar{x}).$$

In most cases we will assume that each code word is equally likely to be sent, that is, $P(\bar{x}) = \frac{1}{M}$. Under this assumption we get

$$P_{\text{ue}}(C, K) = \frac{1}{M} \sum_{\bar{x} \in C} \sum_{\bar{y} \in C \setminus \{\bar{x}\}} P(\bar{y}|\bar{x}).$$

Remark. It is easy to show that for any channel K with additive noise and any coset S of a linear code C we have $P_{\text{ue}}(C, K) = P_{\text{ue}}(S, K)$.

The BSC

To make things more concrete, let us consider as a first (and most important) example the BSC with error probability p. Then

$$P(\bar{y}|\bar{x}) = p^{d_{\text{H}}(\bar{x},\bar{y})}(1-p)^{n-d_{\text{H}}(\bar{x},\bar{y})}$$

since exactly $d_{\text{H}}(\bar{x}, \bar{y})$ bits have to be changed during transmission, the remaining $n - d_{\text{H}}(\bar{x}, \bar{y})$ are unchanged. Hence, if $A_C(z)$ denotes the distance distribution function of C, then

$$\begin{aligned}
P_{\text{ue}}(C, BSC_p) &= \frac{1}{M} \sum_{\bar{x} \in C} \sum_{\bar{y} \in C \setminus \{\bar{x}\}} p^{d_{\text{H}}(\bar{x},\bar{y})}(1-p)^{n-d_{\text{H}}(\bar{x},\bar{y})} \\
&= \sum_{i=1}^{n} A_i(C) p^i (1-p)^{n-i} \\
&= (1-p)^n \sum_{i=1}^{n} A_i(C) \left(\frac{p}{1-p}\right)^i \\
&= (1-p)^n \left\{ A_C\left(\frac{p}{1-p}\right) - 1 \right\}.
\end{aligned}$$

Alternatively, using Lemma 2.2.1, we get

$$A_C\left(\frac{p}{1-p}\right) = \frac{M}{2^n}(1-p)^{-n} A_C^{\text{MW}}(1-2p)$$

and so

$$P_{\text{ue}}(C, BSC_p) = \frac{M}{2^n} A_C^{\text{MW}}(1-2p) - (1-p)^n.$$

We collect these important expressions in a theorem.

Theorem 2.3.1 *Let C be an (n, M) code. Then*

$$\begin{aligned} P_{\text{ue}}(C, BSC_p) &= \sum_{i=1}^{n} A_i(C)p^i(1-p)^{n-i} \\ &= (1-p)^n \left\{ A_C\left(\frac{p}{1-p}\right) - 1 \right\} \\ &= \frac{M}{2^n} A_C^{\text{MW}}(1-2p) - (1-p)^n. \end{aligned}$$

In particular, if C is a linear $[n, k]$ code, then

$$\begin{aligned} P_{\text{ue}}(C, BSC_p) &= 2^{k-n} A_{C^\perp}(1-2p) - (1-p)^n \\ &= 2^{k-n} \sum_{i=0}^{n} A_i(C^\perp)(1-2p)^i - (1-p)^n \\ &= 2^{k-n} \sum_{i=0}^{n} A_i(C^\perp) \left\{ (1-2p)^i - (1-p)^n \right\}. \end{aligned}$$

Example. If C is the $[2^m - 1, 2^m - 1 - m]$ Hamming code, then

$$A_{C^\perp}(z) = 1 + (2^m - 1)z^{2^{m-1}}.$$

Hence

$$\begin{aligned} P_{\text{ue}}(C, BSC_p) &= 2^{-m} \left(1 + (2^m - 1)(1-2p)^{2^{m-1}} \right) - (1-p)^n \\ &= \frac{1}{n+1} \left(1 + n(1-2p)^{\frac{n+1}{2}} \right) - (1-p)^n. \end{aligned}$$

The Binary LBC

Suppose that an $[n, k]$ code C is used on a LBC-1 where the probability that symbol i is in error is given by p_i. A similar argument as for the BSC gives the following expression for the probability of undetected error.

$$\begin{aligned} P_{\text{ue}}(C, p_1, p_2, \cdots, p_n) &= \prod_{i=1}^{n} (1-p_i) \left\{ A_C\left(\frac{p_1}{1-p_1}, \cdots, \frac{p_n}{1-p_n} \right) - 1 \right\} \\ &= 2^{k-n} A_{C^\perp}(1-2p_1, \cdots, 1-2p_n) - \prod_{i=1}^{n} (1-p_i). \end{aligned}$$

Example. Let C be the $[n, n-1]$ even-weight code, that is, $C^\perp = \{\bar{0}, \bar{1}\}$. Then

$$A_{C^\perp}(z_1, z_2, \cdots, z_n) = 1 + z_1 z_2 \cdots z_n,$$

and so

$$P_{\text{ue}}(C, p_1, p_2, \cdots, p_n) = \frac{1}{2}\left\{1 + \prod_{i=1}^{n}(1 - 2p_i)\right\} - \prod_{i=1}^{n}(1 - p_i).$$

The qSC channel

The situation for the qSC is similar to the BSC. First

$$P(\bar{y}|\bar{x}) = \left(\frac{p}{q-1}\right)^{d_H(\bar{x},\bar{y})}(1-p)^{n-d_H(\bar{x},\bar{y})}.$$

An analysis similar to the one above gives the following result.

Theorem 2.3.2 *Let C be an $(n, M; q)$ code. Then*

$$\begin{aligned} P_{\text{ue}}(C, qSC_p) &= \sum_{i=1}^{n} A_i(C)\left(\frac{p}{q-1}\right)^i (1-p)^{n-i} \\ &= (1-p)^n \left\{A_C\left(\frac{p}{(q-1)(1-p)}\right) - 1\right\} \\ &= \frac{M}{q^n} A_C^{\text{MW}}\left(1 - \frac{qp}{q-1}\right) - (1-p)^n. \end{aligned}$$

In particular, if C is a linear $[n, k; q]$ code, then

$$P_{\text{ue}}(C, qSC_p) = q^{k-n} A_{C^\perp}\left(1 - \frac{qp}{q-1}\right) - (1-p)^n.$$

We here also mention another result that has some applications. The result follows directly from 2.2.8.

Theorem 2.3.3 *Let C be an $[n, k; q]$ code and S a proper coset of C. Let $\bar{0}$ be sent over qSC, and let $\bar{y}$ be the received vector. Then*

$$\frac{Pr(\bar{y} \in S)}{Pr(\bar{y} \in C)} \le \frac{1 - \left(1 - \frac{qp}{q-1}\right)^{k+1}}{1 + (q-1)\left(\frac{qp}{q-1}\right)^{k+1}}.$$

The Z-channel

As another concrete example we consider the Z-channel. First we note that if $\bar{x}$ is sent and $\bar{y}$ is received, then necessarily $\bar{y} \leq \bar{x}$. Hence, $P(\bar{y}|\bar{x}) = 0$ if $\bar{y} \not\leq \bar{x}$. If $\bar{y} \leq \bar{x}$, then

$$P(\bar{y}|\bar{x}) = p^{d_H(\bar{x},\bar{y})}(1-p)^{w_H(\bar{x})-d_H(\bar{x},\bar{y})} = p^{w_H(\bar{x})-w_H(\bar{y})}(1-p)^{w_H(\bar{y})}.$$

Therefore

$$P_{\text{ue}}(C, Z_p) = \frac{1}{\#C} \sum_{\bar{x}\in C} \sum_{\substack{\bar{y}\in C \\ \bar{y}<\bar{x}}} p^{w_H(\bar{x})-w_H(\bar{y})}(1-p)^{w_H(\bar{y})}. \tag{2.11}$$

It is interesting to compare $P_{\text{ue}}(C, Z_p)$ and $P_{\text{ue}}(C, BSC_p)$ for some codes. If $\bar{y} \not< \bar{x}$ for all $\bar{y}, \bar{x} \in C$, then clearly $P_{\text{ue}}(C, Z_p) = 0$, whereas $P_{\text{ue}}(C, BSC_p) > 0$ (if $\#C \geq 2$ and $p \in (0,1)$). For other codes, the ratio $P_{\text{ue}}(C, Z_p)/P_{\text{ue}}(C, BSC_p)$ may be large or small, depending on the code and p.

Example. As a simple example, consider the $[2^k-1, k]$ simplex code C_k. This is a linear code where all non-zero code words have Hamming weight 2^{k-1}. Hence $\bar{y} < \bar{x}$ for $\bar{y}, \bar{x} \in C_k$ if and only if $\bar{y} = \bar{0}$ and $\bar{x} \neq \bar{0}$. Therefore

$$P_{\text{ue}}(C_k, Z_p) = \frac{2^k-1}{2^k} p^{2^{k-1}}$$

and

$$P_{\text{ue}}(C_k, BSC_p) = (2^k-1)p^{2^{k-1}}(1-p)^{2^{k-1}-1}.$$

Hence

$$\frac{P_{\text{ue}}(C_k, Z_p)}{P_{\text{ue}}(C_k, BSC_p)} \to \frac{1}{2^k} \quad \text{when} \quad p \to 0,$$

and

$$\frac{P_{\text{ue}}(C_k, Z_p)}{P_{\text{ue}}(C_k, BSC_p)} \to \infty \quad \text{when} \quad p \to 1.$$

Also, for any $p < 1$ we have

$$\frac{P_{\text{ue}}(C_k, Z_p)}{P_{\text{ue}}(C_k, BSC_p)} \to \infty \quad \text{when} \quad k \to \infty.$$

2.4 COMBINED CORRECTION AND DETECTION

In some applications we prefer to use some of the power of a code to correct errors and the remaining power to detect errors. Suppose that C is an $(n, M; q)$ code capable of correcting all error patters with t_0 or less errors that can occur on the channel and suppose that we use the code to correct all error patterns with t errors or less, where $t \leq t_0$. Let $M_t(\bar{x})$ be the set all vectors $\bar{y}$ such that $d_H(\bar{x}, \bar{y}) \leq t$ and such that $\bar{y}$ can be received when $\bar{x}$ is sent over the channel. For two distinct $\bar{x}_1, \bar{x}_2 \in C$, the sets $M_t(\bar{x}_1)$, $M_t(\bar{x}_2)$ are disjoint. If $\bar{y} \in M_t(\bar{x})$ is received, we decode into $\bar{x}$. If $\bar{y} \notin M_t(\bar{x})$ for all $\bar{x} \in C$, then we detect an error.

Suppose that $\bar{x}$ is sent and $\bar{y}$ is received. There are then three possibilities:

1. $\bar{y} \in M_t(\bar{x})$. We then decode, correctly, into $\bar{x}$.
2. $\bar{y} \notin M_t(\bar{x}')$ for all $\bar{x}' \in C$. We then detect an error.
3. $\bar{y} \in M_t(\bar{x}')$ for some $\bar{x}' \in C \setminus \{\bar{x}\}$. We then decode erroneously into $\bar{x}'$, and we have an undetectable error.

Let $P_{\text{ue}}^{(t)} = P_{\text{ue}}^{(t)}(C, K)$ denote the probability that we have an undetectable error. As above we get

$$P_{\text{ue}}^{(t)}(C, K) = \sum_{\bar{x} \in C} P(\bar{x}) \sum_{\bar{x}' \in C \setminus \{\bar{x}\}} \sum_{\bar{y} \in M_t(\bar{x}')} P(\bar{y}|\bar{x}).$$

Assuming that $P(\bar{x}) = \frac{1}{M}$ for all $\bar{x}' \in C$, we get

$$P_{\text{ue}}^{(t)}(C, K) = \frac{1}{M} \sum_{\bar{x} \in C} \sum_{\bar{x}' \in C \setminus \{\bar{x}\}} \sum_{\bar{y} \in M_t(\bar{x}')} P(\bar{y}|\bar{x}).$$

The BSC

Consider as an example the BSC with error probability p. For the BSC we have

$$M_t(\bar{x}) = \{\bar{y} \mid d_H(\bar{x}, \bar{y}) \leq t\}.$$

Let C be an (n, M) code of minimum distance $d \geq 2t + 1$. As for pure error detection we get

$$P_{\text{ue}}^{(t)}(C, BSC_p) \quad = \quad (1-p)^n \frac{1}{M} \sum_{\bar{x} \in C} \sum_{\bar{x}' \in C \setminus \{\bar{x}\}} \sum_{\bar{y} \in M_t(\bar{x}')} \left(\frac{p}{1-p}\right)^{d_H(\bar{x}, \bar{y})}$$

$$= \quad (1-p)^n \left\{ A_C^{(t)} \left(\frac{p}{1-p} \right) \right\} \tag{2.12}$$

where

$$A_C^{(t)}(z) = \sum_{j=d-t}^{n} A_{t,j} z^j,$$

and

$$A_{t,j} = \sum_{i=j-t}^{j+t} A_i N_t(i,j)$$

is the number of vectors of weight j within distance t of some non-zero code word, and $N_t(i,j)$ was defined by (2.10).

2.5 COMMENTS AND REFERENCES

2.1 Most of this material can be found in most text books on error-correcting codes, see the general bibliography. However, many of the books restrict themselves to binary codes.

The weight hierarchy (under a different name) was first studied by Helleseth, Kløve and Mykkeltveit [80]. The r-th minimum support weight is also known as *r-th generalized Hamming weight* [200]. The inequality 2.1 was shown by Helleseth, Kløve and Ytrehus in [81] (for $q = 2$) and [82] (for general q).

For the referred standard CRC codes, see A. S. Tanenbaum: Computer Networks, 2ed. Prentice Hall Publ., Englewood Cliffs, NJ, 1989, and Castagnoli et al. [45].

2.2. Much of this material is also standard. Theorem 2.2.2 is due to MacWilliams [142]. Theorem 2.2.1 is due to Delsarte [53]. Proof of both (for $q = 2$) can be found in MacWilliams & Sloane [13, Chapter 5]. Theorem 2.2.4 (for $q = 2$) was given by Kasami et al. [93].

Theorem 2.2.5 is from Kløve [103].

Theorem 2.2.6 and Corollary 2.2.2 are due to Assmus and Mattson [28]. Theorem 2.2.7 is essentially due to Ancheta [27].

Theorem 2.2.8 with $q = 2$ is due to Sullivan [183]. An alternative proof and generalization to general q was given by Redinbo [167]. Further results are given in Kløve [104], [105], [106].

We remark that the weight distribution of cosets can be useful in the wire-tap channel area, see Wyner [210] and Korzhik and Yakovlev [121].

Theorem 2.2.10 is essentially due to MacWilliams [142]. In the present form it was given in Kløve [99].

Theorem 2.2.11 is new here. Special cases were given implicitly in Korzhik and Fink [10] and Kasami et al. [94].

2.3. Theorem 2.3.1 is standard, and may be found in many references, even if the notation may be different. See e.g. [11].

2.4. A more detailed discussion is found in Kløve [99].

3

ERROR DETECTING CODES FOR THE BSC

The binary symmetric channel is central in many applications and we will therefore give a fairly complete account of the known results. Many of the results can be generalized to the qSC in a straight forward manner.

3.1 BASIC FORMULAS AND BOUNDS

In Theorem 2.3.1 we gave formulas for $P_{\mathrm{ue}}(C, BSC_p)$. To simplify the notation slightly, we will write $P_{\mathrm{ue}}(C,p)$, or just $P_{\mathrm{ue}}(p)$, for $P_{\mathrm{ue}}(C, BSC_p)$ in this chapter.

Define

$$\begin{aligned} P_{\mathrm{ue}}(n,M,p) &= \min\{P_{\mathrm{ue}}(C,p) \mid C \text{ is an } (n,M) \text{ code}\}, \\ P_{\mathrm{ue}}[n,k,p] &= \min\{P_{\mathrm{ue}}(C,p) \mid C \text{ is an } [n,k] \text{ code}\}. \end{aligned}$$

An (n,M) code C is called an *optimal general code* for p if $P_{\mathrm{ue}}(C,p) = P_{\mathrm{ue}}(n,M,p)$. Similarly, an $[n,k]$ is called an *optimal linear code* for p if $P_{\mathrm{ue}}(C,p) = P_{\mathrm{ue}}[n,k,p]$. Note that a linear code may be optimal linear without being optimal general. However, to simplify the language, we usually talk about *optimal codes,* meaning optimal general if the code is non-linear and optimal linear if the code is linear.

It turns out that $P_{\mathrm{ue}}(n,M,p)$ decreases exponentially with n. Therefore, define

$$\pi_{\mathrm{ue}}(n,R,p) = -\frac{\log_2 P_{\mathrm{ue}}(n, \lceil 2^{Rn} \rceil, p)}{n},$$

$$\overline{\pi}_{\text{ue}}(R,p) = \limsup_{n\to\infty} \pi_{\text{ue}}(n,R,p),$$

$$\underline{\pi}_{\text{ue}}(R,p) = \liminf_{n\to\infty} \pi_{\text{ue}}(n,R,p),$$

and

$$\pi_{\text{ue}}(R,p) = \lim_{n\to\infty} \pi_{\text{ue}}(n,R,p)$$

when the limit exists. A lower (resp. upper) bound on $P_{\text{ue}}(n,M)$ will give an upper (resp. lower) bound on $\pi_{\text{ue}}(n,R,p)$.

We note that $P_{\text{ue}}\left(\frac{1}{2}\right) = \frac{2^k-1}{2^n}$. It used to be believed that since $p = \frac{1}{2}$ is the "worst case", it would be true that $P_{\text{ue}}(p) \leq \frac{2^k-1}{2^n}$ for all $p \in [0, \frac{1}{2}]$. However, this is not the case. A very simple example is given by the $[3,2]$ code $C = \{000, 100, 010, 110\}$. For this code

$$P_{\text{ue}}(p) = 2p(1-p)^2 + p^2(1-p),$$

this is plotted in Figure 3.1. For example,

$$P_{\text{ue}}(0.4) = 0.384 > 0.375 = P_{\text{ue}}(0.5).$$

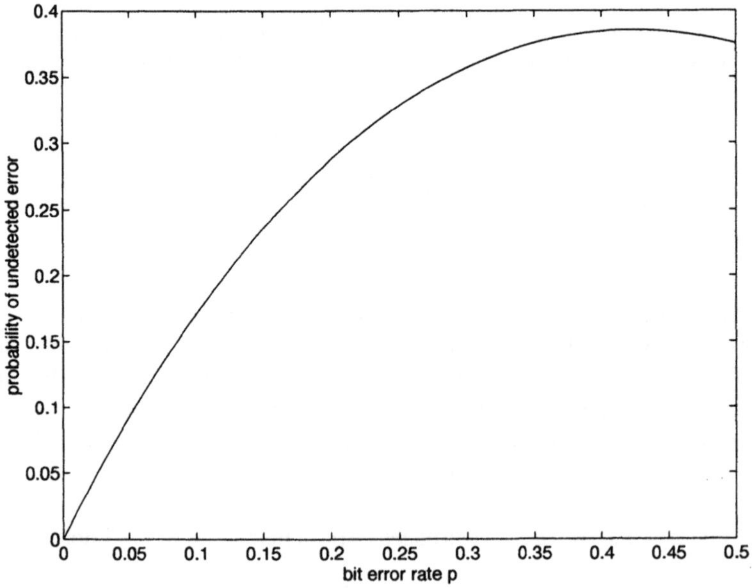

Figure 3.1 Plot of $P_{\text{ue}}(p)$ for the code $C = \{000, 100, 010, 110\}$.

A code C is called ***good*** if

$$P_{ue}(C,p) \leq P_{ue}\left(C, \frac{1}{2}\right) = \frac{2^k - 1}{2^n} \tag{3.1}$$

for all $p \in [0, \frac{1}{2}]$, and it is called ***proper*** if $P_{ue}(C,p)$ is monotonously increasing on $[0, \frac{1}{2}]$.

When we want to find an (n, M) or $[n, k]$ code for error detection in some application, the best choice is an optimal code, ***provided*** we know p. There are two problems. First, we may not know p, and a code optimal for $p' \neq p$ may not be optimal for p. Moreover, even if we know p, there is in general no method to find an optimal code, except exhaustive search, and this is in most cases not feasible. Therefore, it is useful to have some criterion by which we can judge the usefulness of a given code for error detection. An engineering rule of thumb is that if a code is good in the sense just defined, then it is good enough for most practical applications. It has not been proved that there exist good $[n, k]$ codes for all n and k, but numerical evidence indicate that this is the case. We shall later show that a number of well known classes of codes are good. On the other hand, many codes are not good. Therefore, it is important to have methods to decide if a code is good or not.

The bound $\frac{2^k-1}{2^n}$ in the definition of a good code is to some extent arbitrary. Another bound that is often used is 2^{k-n}. For most practical applications, any bound of the same order of magnitude would do. Let $\mathcal{C}$ be an infinite class of codes. We say that $\mathcal{C}$ is ***asymptotically good*** if there exists a constant c such that

$$P_{ue}(C,p) \leq cP_{ue}\left(C, \frac{1}{2}\right)$$

for all $C \in \mathcal{C}$ and all $p \in \left[0, \frac{1}{2}\right]$. Otherwise we say that $\mathcal{C}$ is asymptotically bad.

Clearly, if a code is proper, then it is good. As a first illustration, we show that Hamming codes are proper. As shown in Section 2.3, for the Hamming code we have

$$P_{ue}(p) = \frac{1}{n+1}\left(1 + n(1-2p)^{\frac{n+1}{2}}\right) - (1-p)^n.$$

Hence for $p \in \left(0, \frac{1}{2}\right]$ we get

$$\frac{d}{dp}P_{ue}(p) = n\left\{\left((1-p)^2\right)^{\frac{n-1}{2}} - (1-2p)^{\frac{n-1}{2}}\right\} > 0.$$

We will give a couple of necessary conditions for a code to be good.

Theorem 3.1.1 *If C is a good binary linear $[n, k, d]$ code, then*

$$1 - \frac{k}{n} < H_2\left(\frac{d}{n}\right).$$

Proof: Choosing $p = \frac{d}{n}$ we get

$$2^{k-n} > \frac{2^k - 1}{2^n} \geq \left(\frac{d}{n}\right)^d \left(1 - \frac{d}{n}\right)^{n-d},$$

and so

$$k - n > d \log_2\left(\frac{d}{n}\right) + (n-d)\log_2\left(1 - \frac{d}{n}\right) = -nH_2\left(\frac{d}{n}\right).$$

Remark. For $\frac{d}{n} \leq \frac{1}{2}$ the condition of Theorem 3.1.1 can alternatively be written as $\rho\left(\frac{k}{n}\right) \leq \frac{d}{n}$. Asymptotically, it means that a long good code must satisfy the Gilbert-Varshamov bound.

Theorem 3.1.2 *Let C be a good binary linear $[n, k]$ code. Let its weight distribution be $A_0, A_1, \cdots, A_n$ and let the weight distribution of the dual code be $B_0, B_1, \cdots, B_n$. Then, for all i, $0 < i < \frac{n}{2}$ we have*

$$\begin{aligned} A_i &\leq \frac{2^k - 1}{2^n}\left(\frac{n}{i}\right)^i \left(\frac{n}{n-i}\right)^{n-i}, \\ B_i &< 2^{-k}\left(\frac{n}{i}\right)^i \left(\frac{n}{n-i}\right)^{n-i}. \end{aligned}$$

Proof: Choosing $p = \frac{i}{n}$ we get

$$\frac{2^k - 1}{2^n} \geq P_{\text{ue}}\left(C, \frac{i}{n}\right) = \sum_{j=1}^{n} A_j \left(\frac{i}{n}\right)^j \left(\frac{n-i}{n}\right)^{n-j} \geq A_i \left(\frac{i}{n}\right)^i \left(\frac{n-i}{n}\right)^{n-i}.$$

For the other inequality, we first note that

$$2^{k-n} > \frac{2^k - 1}{2^n} \geq P_{\text{ue}}(C, p) = 2^{k-n} + 2^{k-n}\sum_{j=1}^{n} B_j (1-2p)^j - (1-p)^n$$

and so

$$B_i(1-2p)^i \le \sum_{j=1}^{n} B_j(1-2p)^j \le 2^{n-k}(1-p)^n.$$

Choosing $1-2p = \frac{i}{n-i}$ we get

$$B_i\left(\frac{i}{n-i}\right)^i < 2^{-k}\left(\frac{n}{n-i}\right)^n.$$

Theorem 3.1.3 *Let C be a good binary linear $[n,k]$ code. Let the weight distribution of the dual code be $B_0, B_1, \cdots, B_n$. Then*

$$B_1 2^k \le n.$$

Proof: From $P_{\mathrm{ue}}(C,p) = 2^{k-n}\sum_{i=0}^{n} B_i(1-2p)^i - (1-p)^n$ we get

$$\frac{d}{dp}P_{\mathrm{ue}}(C,p) = n(1-p)^{n-1} - 2^{k-n}\sum_{i=1}^{n} 2iB_i(1-2p)^{i-1}.$$

Since $P_{\mathrm{ue}}(C,p)$ can not be decreasing at $p=\frac{1}{2}$ for a good code, we get

$$\frac{d}{dp}P_{\mathrm{ue}}(C,p)\Big|_{p=\frac{1}{2}} = \left(\frac{1}{2}\right)^{n-1}\{n - B_1 2^k\} \ge 0.$$

For a practical application we may know that $p \le p_0$ for some fixed p_0. If we use an $[n,k,d]$ with $d \ge p_0 n$, then the next theorem shows that $P_{\mathrm{ue}}(p) \le P_{\mathrm{ue}}(p_0)$ for all $p \le p_0$.

Theorem 3.1.4 *Let C be an $[n,k,d]$ code. Then $P_{\mathrm{ue}}(C,p)$ is monotonously increasing on $\left[0, \frac{d}{n}\right]$.*

Proof: Since $p^i(1-p)^{n-i}$ is monotonously increasing on $\left[0, \frac{i}{n}\right]$, and in particular on $\left[0, \frac{d}{n}\right]$ for all $i \ge d$, the theorem follows.

How to determine if a polynomial has a zero

In a number of cases we want to know if a polynomial has a zero in a given interval. For example, if $\frac{d}{dp}P_{\mathrm{ue}}(C,p) > 0$ for all $p \in \left(0, \frac{1}{2}\right)$, then C is proper. If

$P_{ue}(C,p) \le P_{ue}\left(C,\frac{1}{2}\right)$ then C is good (by definition). If $A_{C_1}(z) - A_{C_2}(z) > 0$ for $z \in (0,1)$, then $P_{ue}(C_1,p) > P_{ue}(C_2,p)$ for all $p \in \left(0,\frac{1}{2}\right)$, and so C_2 is better for error detection then C_1. The simplest way is to use some mathematical software to draw the graph and inspect it and/or calculate the roots. However, this may sometimes be misleading. Therefore, we give a short description of the systematic method of Sturm sequences.

Let $f(z)$ be a polynomial and $[a,b]$ an interval.

Define the sequence $f_0(z), f_1(z), \cdots, f_m(z)$ by the Euclidean algorithm:

$f_0(z) = f(z)$,
$f_1(z) = f'(z)$,
$f_i(z) \equiv -f_{i-2}(z) \pmod{f_{i-1}(z)}$, $\deg(f_i(z)) < \deg(f_{i-1}(z))$ for $2 \le i \le m$,
$f_m(z)$ divides $f_{m-1}(z)$.

Define $g_0(z), g_1(z), \cdots, g_m(z)$ by

$$g_i(z) = \frac{f_i(z)}{f_m(z)}.$$

If $g_0(a) = 0$ we divide all $g_i(z)$ with $(z-a)$; and if $g_0(b) = 0$ we divide all $g_i(z)$ with $(z-b)$.

Lemma 3.1.1 *The number of distinct zeros for $f(z)$ in (a,b) is given by*

$$\#\{i \mid g_{i-1}(a)g_i(a) < 0,\ 1 \le i \le m\} - \#\{i \mid g_{i-1}(b)g_i(b) < 0,\ 1 \le i \le m\}.$$

Another method which is not guaranteed to work, but is simpler when it does, is to rewrite $f(z)$ into the form

$$f(z) = \sum_{i=0}^{n} a_i (z-a)^i (b-z)^{n-i}$$

where $n = \deg(f(z))$. If $a_i \ge 0$ for all i, then clearly $f(z) \ge 0$ for all $z \in [a,b]$.

We illustrate this last method by giving some results that are also of theoretical interest.

Let C be an $[n,k,d]$ code with a dual $[n,n-k,d']$ code. We have

$$\begin{aligned}
P_{\text{ue}}(C,p) &= \sum_{i=1}^{n} A_i(C)p^i(1-p)^{n-i} \\
&= \sum_{i=1}^{n} A_i(C)p^i(p+1-2p)^{n-i} \\
&= \sum_{i=1}^{n} A_i(C)p^i \sum_{j=0}^{n-i} \binom{n-i}{j} p^j(1-2p)^{n-i-j} \\
&= \sum_{l=1}^{n} \alpha_l p^l(1-2p)^{n-l},
\end{aligned}$$

where

$$\alpha_l = \sum_{i=1}^{l} A_i(C)\binom{n-i}{l-i}.$$

In particular, $\alpha_l = 0$ for $1 \le l < d$. Similarly,

$$\begin{aligned}
P_{\text{ue}}(C,p) &= 2^{k-n}\sum_{i=0}^{n} A_i(C^\perp)(1-2p)^i - (1-p)^n \\
&= 2^{k-n}\sum_{i=0}^{n} A_i(C^\perp)(1-2p)^i(2p+1-2p)^{n-i} - (p+1-2p)^n \\
&= 2^{k-n}\sum_{i=0}^{n} A_i(C^\perp)\sum_{j=0}^{n-i}\binom{n-i}{j}2^jp^j(1-2p)^{n-j} \\
&\quad -\sum_{j=0}^{n}\binom{n}{j}p^j(1-2p)^{n-j}.
\end{aligned}$$

Hence, an alternative expression for α_l is the following.

$$\alpha_l = 2^{l+k-n}\sum_{i=0}^{n-l} A_i(C^\perp)\binom{n-i}{l} - \binom{n}{l}.$$

Further,

$$P_{\text{ue}}\left(C,\frac{1}{2}\right) = \frac{2^k-1}{2^n}(2p+1-2p)^n = \frac{2^k-1}{2^n}\sum_{l=0}^{n}\binom{n}{l}2^lp^l(1-2p)^{n-l}.$$

Therefore, if

$$\alpha_l \leq \frac{2^l(2^k-1)}{2^n}\binom{n}{l} \tag{3.2}$$

for $1 \leq l \leq n$, then C is good. For $1 \leq l < d$ we have $\alpha_l = 0$ so (3.2) is satisfied. For $0 \leq n-l < d'$ we have

$$\alpha_l = 2^{l+k-n}\binom{n}{l} - \binom{n}{l} = \left\{\frac{2^l(2^k-1)}{2^n} - (1-2^{l-n})\right\}\binom{n}{l}$$

and so (3.2) is again satisfied. Therefore we get the following theorem.

Theorem 3.1.5 *Let C be an $[n,k,d]$ code and let d' be the minimum distance of the dual code. If*

$$\sum_{i=1}^{n} A_i(C)\binom{n-i}{l-i} \leq \frac{2^l(2^k-1)}{2^n}\binom{n}{l}2^l$$

for $d \leq l \leq n-d'$, then C is good.

As a further example, we show that the $[2^m, 2^m-m-1, 4]$ extended Hamming code is proper. For convenience we write $M = 2^{m-1}$. Then

$$P_{\mathrm{ue}}(p) = \frac{1}{4M}\left\{1 + (4M-2)(1-2p)^M + (1-2p)^{2M}\right\} - (1-p)^{2M},$$

and so

$$\begin{aligned}
\frac{d}{dp}P_{\mathrm{ue}}(p) &= 2M(1-p)^{2M-1} - (2M-1)(1-2p)^{M-1} - (1-2p)^{2M-1} \\
&= 2M\sum_{i=0}^{2M-1}\binom{2M-1}{i}p^i(1-2p)^{2M-1-i} \\
&\quad -(2M-1)\sum_{i=0}^{M}\binom{M}{i}2^ip^i(1-2p)^{2M-1-i} - (1-2p)^{2M-1} \\
&= \sum_{i=0}^{2M-1}\alpha_i p^i(1-2p)^{2M-1-i},
\end{aligned}$$

where $\alpha_0 = \alpha_1 = \alpha_2 = 0$, and

$$i!\alpha_i = 2M(2M-1)(2M-2)\left\{\prod_{j=3}^{i}(2M-j) - \prod_{j=3}^{i}(2M-2j+2)\right\} > 0$$

for $3 \leq i \leq 2M-1$. Hence $\frac{d}{dp}P_{\mathrm{ue}}(p) \geq 0$ for all $p \in \left[0, \frac{1}{2}\right]$; that is, C is proper.

A Sufficient Condition

For linear codes with code words of even weight only, the following theorem is sometimes useful to show that

$$P_{\mathrm{ue}}(C,p) \le 2^{k-n}\left\{1+(1-2p)^n - 2(1-p)^n\right\}.$$

Note that $1+(1-2p)^n - 2(1-p)^n$ goes monotonically to 0 when p goes to 0.

Since all code words have even weight, the all-one vector $\bar{1}$ belongs to $C^\perp$. Therefore, for any code word $\bar{x} \in C^\perp$ of weight i, there is a code word in $C^\perp$ of weight $n-i$, namely $\bar{x}+\bar{1}$. Hence we can write the weight distribution function of $C^\perp$ as follows:

$$A_{C^\perp}(z) = \sum_{i=0}^{\nu} B_i(z^i + z^{n-i}),$$

where $\nu = \left\lfloor \frac{n}{2} \right\rfloor$.

Remark. If n is even, then $B_\nu = \frac{1}{2}A_\nu(C^\perp)$. For $i < \frac{n}{2}$ we have $B_i = A_i(C^\perp)$.

Theorem 3.1.6 *Let C be an $[n,k,d]$ code where $n = 2\nu+1$ is odd and all code words have even weight. Let d' be the minimum distance of the dual code. If*

$$(2^{n-k}-2)\binom{\nu}{j} \ge 2^{2j+1}\sum_{i=j}^{\nu-1}\binom{i+j}{2j}B_{\nu-i}$$

for

$$\frac{d}{2} \le j < \frac{(n-2d')^2+n}{2n},$$

then

$$P_{\mathrm{ue}}(C,p) \le 2^{k-n}\left\{1+(1-2p)^n - 2(1-p)^n\right\}$$

for all $p \in \left[0, \frac{1}{2}\right]$.

Theorem 3.1.7 *Let C be an $[n,k,d]$ code where $n = 2\nu$ is even and all code words have even weight. Let d' be the minimum distance of the dual code. If*

$$(2^{n-k}-2)\binom{\nu}{j} \ge 2^{2j}\sum_{i=j}^{\nu-1}\left\{\binom{i+j}{2j}+\binom{i+j-1}{2j}\right\}B_{\nu-i}$$

for

$$\frac{d}{2} \le j < \frac{(n-2d')^2+n-2}{2n-2},$$

then

$$P_{\mathrm{ue}}(C,p) \le 2^{k-n}\left\{(1+(1-2p)^n - 2(1-p)^n\right\}$$

for all $p \in \left[0, \frac{1}{2}\right]$.

We sketch the proof of Theorem 3.1.6. By Theorem 2.3.1

$$\begin{aligned} P_{\mathrm{ue}}(C,p) &= 2^{n-k} A_{C^\perp}(1-2p) - (1-p)^n \\ &= 2^{k-n}\left\{1+(1-2p)^n - 2(1-p)^n - F(p)\right\} \end{aligned}$$

where

$$F(p) = (2^{n-k}-2)(1-p)^n - \sum_{i=1}^{\nu} B_i\left\{(1-2p)^i + (1-2p)^{n-i}\right\}.$$

We will show that $F(p) \ge 0$ for all p. We see that

$$(1-p)^n = (1-p)\sum_{j=0}^{\nu}\binom{\nu}{i} p^{2j}(1-2p)^{\nu-j}$$

and

$$(1-2p)^i + (1-2p)^{n-i} = (1-p)\sum_{j=0}^{i} 2^{2j+1}\binom{i+j}{2j} p^{2j}(1-2p)^{\nu-j}.$$

Hence

$$F(p) = (1-p)\sum_{j=0}^{\nu} p^{2j}(1-2p)^{\nu-j}F_j,$$

where

$$F_j = (2^{n-k}-2)\binom{\nu}{j} - \sum_{i=j}^{\nu-1} 2^{2j+1}\binom{i+j}{2j} B_{\nu-i}.$$

We will show that $F_j \ge 0$ for all j. By assumption, this is true for $\frac{d}{2} \le j < d_0 \stackrel{\mathrm{def}}{=} \left\lceil \frac{(n-2d')^2+n}{2n} \right\rceil$. Consider $j < \frac{d}{2}$. Then

$$\sum_{i=j}^{\nu-1} 2^{2j+1}\binom{i+j}{2j} B_{\nu-i} \quad = \quad 2^{2j}\sum_{i=0}^{n} B_i \binom{i+j+\nu-n}{2j} - 2^{2j+1}\binom{j+\nu}{2j}$$

$$\begin{aligned} &= 2^{2j}\sum_{l=0}^{2j}\binom{j+\nu-n}{2j-l}\sum_{i=0}^{n}B_i\binom{i}{l} - 2^{2j+1}\binom{j+\nu}{2j} \\ &= 2^{2j}\sum_{l=0}^{2j}\binom{j+\nu-n}{2j-l}2^{n-k-l}\binom{n}{l} - 2^{2j+1}\binom{j+\nu}{2j} \\ &= 2^{n-k}\binom{\nu}{j} - 2^{2j+1}\binom{j+\nu}{2j}, \end{aligned}$$

where the second to last equality follows from Theorem 2.2.3, and so

$$F_j = 2^{2j+1}\binom{j+\nu}{2j} - 2\binom{\nu}{j} \geq 0.$$

Finally, consider $j \geq d_0$. We prove that $F_j \geq 0$ by induction on j. As basis for the induction, we use that $F_{d_0-1} \geq 0$. We observe that $B_{\nu-i} = 0$ for $n - d' < i \leq \nu - 1$. Further

$$\begin{aligned} \frac{2^{2j+1}\binom{i+j}{2j}}{\binom{\nu}{j}} &= \frac{2(i+j)(i-j+1)}{(2j-1)(\nu-j+1)}\frac{2^{2(j-1)j+1}\binom{i+(j-1)j}{2(j-1)j}}{\binom{\nu}{j-1}} \\ &\leq \frac{2^{2(j-1)j+1}\binom{i+(j-1)j}{2(j-1)j}}{\binom{\nu}{j-1}} \end{aligned}$$

for $j \geq d_0$ and $j \leq i \leq \nu - d'$. Hence

$$\begin{aligned} \frac{F_j}{\binom{\nu}{j}} &= (2^{n-k}-2) - \sum_{i=j}^{\nu-d'} B_{\nu-i}\frac{2^{2j+1}\binom{i+j}{2j}}{\binom{\nu}{j}} \\ &\geq (2^{n-k}-2) - \sum_{i=j-1}^{\nu-d'} B_{\nu-i}\frac{2^{2(j-1)j+1}\binom{i+(j-1)j}{2(j-1)j}}{\binom{\nu}{j-1}} \\ &= \frac{F_{j-1}}{\binom{\nu}{j-1}} \geq 0 \end{aligned}$$

by the induction hypothesis.

Remark. If

$$\left\lceil \frac{d}{2} \right\rceil \geq \frac{(n-2d')^2+n}{2n},$$

then there are no further values of j to check and similarly for n even. Hence we get the following corollary.

Corollary 3.1.1 *Let C be an $[n,k,d]$ code where all code words have even weight. Let d' be the minimum distance of the dual code. If n is odd and*

$$\left\lceil \frac{d}{2} \right\rceil \geq \frac{(n-2d')^2+n}{2n}$$

or n is even and

$$\left\lceil \frac{d}{2} \right\rceil \geq \frac{(n-2d')^2+n-2}{2n-2}$$

then

$$P_{\mathrm{ue}}(C,p) \leq 2^{k-n}\left\{1+(1-2p)^n-2(1-p)^n\right\}$$

for all $p \in \left[0, \frac{1}{2}\right]$.

3.2 FINDING THE AVERAGE

In this section we consider the mean probability of undetected error for the codes in some set $\mathcal{C}$ of linear codes of length n.

The common notations for the *mean* and *variance* of a stochastic variable X are $E(X)$ and $Var(X) = E(X^2) - E(X)^2$. Further $\sigma = \sqrt{Var(X)}$, the *standard deviation* of X.

For our particular application use the notations

$$P_{\mathrm{ue}}(\mathcal{C},p) = \frac{1}{\#\mathcal{C}} \sum_{C\in\mathcal{C}} P_{\mathrm{ue}}(C,p),$$

and

$$Var(\mathcal{C},p) = Var(\{P_{\mathrm{ue}}(C,p) \mid C \in \mathcal{C}\}).$$

Further, let

$$A_i(\mathcal{C}) = \frac{1}{\#\mathcal{C}} \sum_{C\in\mathcal{C}} A_i(C).$$

and for $\bar{x}, \bar{y} \in GF(2)^n$, let

$$\alpha(\bar{x}) = \alpha_{\mathcal{C}}(\bar{x}) = \frac{\#\{C \in \mathcal{C} \mid \bar{x} \in C\}}{\#\mathcal{C}},$$

$$\alpha(\bar{x},\bar{y}) = \alpha_{\mathcal{C}}(\bar{x},\bar{y}) = \frac{\#\{C \in \mathcal{C} \mid \bar{x},\bar{y} \in C\}}{\#\mathcal{C}}.$$

From the definitions and Theorem 2.3.1 we get the following result.

Theorem 3.2.1 *Let $\mathcal{C}$ be a set of linear codes of length n. Then*

$$P_{\text{ue}}(\mathcal{C},p) = \sum_{i=1}^{n} A_i(\mathcal{C}) p^i (1-p)^{n-i}.$$

Lemma 3.2.1 *Let $\mathcal{C}$ be a set of linear codes of length n. Then*

$$A_i(\mathcal{C}) = \sum_{\substack{\bar{x}\in GF(2)^n \\ w_{\mathrm{H}}(\bar{x})=i}} \alpha(\bar{x}).$$

Proof: We have

$$\begin{aligned} A_i(\mathcal{C}) &= \frac{1}{\#\mathcal{C}} \sum_{C\in\mathcal{C}} \sum_{\substack{\bar{x}\in C \\ w_{\mathrm{H}}(\bar{x})=i}} 1 \\ &= \sum_{\substack{\bar{x}\in GF(2)^n \\ w_{\mathrm{H}}(\bar{x})=i}} \frac{1}{\#\mathcal{C}} \sum_{\substack{C\in\mathcal{C} \\ \bar{x}\in C}} 1 \\ &= \sum_{\substack{\bar{x}\in GF(2)^n \\ w_{\mathrm{H}}(\bar{x})=i}} \alpha(\bar{x}). \end{aligned}$$

Theorem 3.2.2 *Let $\mathcal{C}$ be a set of linear codes of length n. Then*

$$Var(\mathcal{C},p) = \sum_{\bar{x},\bar{y}\in GF(2)^n\setminus\{\bar{0}\}} \alpha(\bar{x},\bar{y}) p^{w_H(\bar{x})+w_H(\bar{y})} (1-p)^{2n-w_H(\bar{x})-w_H(\bar{y})} - P_{\text{ue}}(\mathcal{C},p)^2.$$

Proof: We have

$$E(P_{\text{ue}}(p)^2) = \frac{1}{\#\mathcal{C}} \sum_{C\in\mathcal{C}} P_{\text{ue}}(C,p)^2$$

$$\begin{aligned}
&= \frac{1}{\#\mathcal{C}} \sum_{C \in \mathcal{C}} \sum_{\bar{x} \in C \setminus \{\bar{0}\}} p^{w_H(\bar{x})}(1-p)^{n-w_H(\bar{x})} \\
&\qquad \sum_{\bar{y} \in C \setminus \{\bar{0}\}} p^{w_H(\bar{y})}(1-p)^{n-w_H(\bar{y})} \\
&= \sum_{\bar{x},\bar{y} \in GF(2)^n \setminus \{\bar{0}\}} \alpha(\bar{x},\bar{y}) p^{w_H(\bar{x})+w_H(\bar{y})}(1-p)^{2n-w_H(\bar{x})-w_H(\bar{y})}.
\end{aligned}$$

Next, we consider the mean and the variance for some special sets of codes. For some fixed $[n,\kappa]$ code K, let

$$\mathcal{C}_k(K) = \{C \mid C \subseteq K \quad \text{and} \quad dim(C) = k\}$$

denote the set of $[n,k]$ subcodes of K.

Theorem 3.2.3 *Let K be an $[n,\kappa]$ code and $k \leq \kappa$. Then*

$$P_{\text{ue}}(\mathcal{C}_k(K),p) = \frac{2^k-1}{2^\kappa-1} P_{\text{ue}}(K,p),$$

$$\begin{aligned}
Var(\mathcal{C}_k(K),p) &= \frac{(2^k-1)(2^\kappa-2^k)}{(2^\kappa-1)(2^\kappa-2)}(1-p)^{2n}\left\{A_K\left(\frac{p^2}{(1-p)^2}\right)-1\right\} \\
&\quad -\frac{(2^k-1)(2^\kappa-2^k)}{(2^\kappa-1)^2(2^\kappa-2)} P_{\text{ue}}(K,p)^2.
\end{aligned}$$

Proof: First,

$$\#\mathcal{C}_k(K) = \begin{bmatrix}\kappa \\ k\end{bmatrix} = \frac{\prod_{i=0}^{k-1}(2^\kappa-2^i)}{\prod_{i=0}^{k-1}(2^k-2^i)}.$$

We see that if $\bar{x} \notin K$, then $\alpha(\bar{x}) = 0$. On the other hand, if $\bar{x} \in K$, then

$$\#\{C \in \mathcal{C}_k(K) \mid \bar{x} \in C\} = \begin{bmatrix}\kappa-1 \\ k-1\end{bmatrix},$$

and so

$$\alpha(\bar{x}) = \frac{\begin{bmatrix}\kappa-1 \\ k-1\end{bmatrix}}{\begin{bmatrix}\kappa \\ k\end{bmatrix}} = \frac{2^k-1}{2^\kappa-1},$$

and

$$A_i(\mathcal{C}_k(K)) = \frac{2^k-1}{2^\kappa-1} A_i(K). \tag{3.3}$$

Hence,

$$P_{\mathrm{ue}}(\mathcal{C}_k(K),p)=\frac{2^k-1}{2^\kappa-1}P_{\mathrm{ue}}(K,p).$$

Similarly, if $\bar{x}\notin K$ or $\bar{y}\notin K$, then

$$\alpha(\bar{x},\bar{y})=0;$$

if $\bar{x},\bar{y}\in K\setminus\{\bar{0}\}$ and $\bar{x}\neq\bar{y}$, then

$$\alpha(\bar{x},\bar{y})=\frac{(2^k-1)(2^k-2)}{(2^\kappa-1)(2^\kappa-2)};$$

if $\bar{x}=\bar{y}$ and $\bar{x}\in K\setminus\{\bar{0}\}$, then

$$\alpha(\bar{x},\bar{y})=\alpha(\bar{x})=\frac{2^k-1}{2^\kappa-1}.$$

Putting these values into Theorem 3.2.2 we get

$$\begin{aligned}
&Var(\mathcal{C}_k(K),p)\\
&= \frac{(2^k-1)(2^k-2)}{(2^\kappa-1)(2^\kappa-2)}\sum_{\bar{x}\in K\setminus\{\bar{0}\}}p^{w_H(\bar{x})}(1-p)^{n-w_H(\bar{x})}\\
&\qquad\sum_{\bar{y}\in K\setminus\{\bar{0},\bar{x}\}}p^{w_H(\bar{y})}(1-p)^{n-w_H(\bar{y})}\\
&\quad+\frac{2^k-1}{2^\kappa-1}\sum_{\bar{x}\in K\setminus\{\bar{0}\}}p^{2w_H(\bar{x})}(1-p)^{2n-2w_H(\bar{x})}-\left\{\frac{2^k-1}{2^\kappa-1}P_{\mathrm{ue}}(K,p)\right\}^2\\
&= \frac{(2^k-1)(2^k-2)}{(2^\kappa-1)(2^\kappa-2)}\sum_{\bar{x}\in K\setminus\{\bar{0}\}}p^{w_H(\bar{x})}(1-p)^{n-w_H(\bar{x})}P_{\mathrm{ue}}(K,p)\\
&\quad-\frac{(2^k-1)(2^k-2)}{(2^\kappa-1)(2^\kappa-2)}\sum_{\bar{x}\in K\setminus\{\bar{0}\}}p^{2w_H(\bar{x})}(1-p)^{2n-2w_H(\bar{x})}\\
&\quad+\frac{2^k-1}{2^\kappa-1}\sum_{\bar{x}\in K\setminus\{\bar{0}\}}p^{2w_H(\bar{x})}(1-p)^{2n-2w_H(\bar{x})}-\left\{\frac{2^k-1}{2^\kappa-1}P_{\mathrm{ue}}(K,p)\right\}^2\\
&= \left\{\frac{(2^k-1)(2^k-2)}{(2^\kappa-1)(2^\kappa-2)}-\frac{(2^k-1)^2}{(2^\kappa-1)^2}\right\}P_{\mathrm{ue}}(K,p)^2\\
&\quad-\left\{\frac{(2^k-1)(2^k-2)}{(2^\kappa-1)(2^\kappa-2)}-\frac{2^k-1}{2^\kappa-1}\right\}(1-p)^{2n}\left\{A_K\left(\frac{p^2}{(1-p)^2}\right)-1\right\}.
\end{aligned}$$

Choosing $K=GF(2)^n$ we get the following corollaries.

Corollary 3.2.1 *We have*

$$\begin{aligned} P_{\text{ue}}(\mathcal{C}_k(GF(2)^n),p) &= \frac{2^k-1}{2^n-1}\left\{1-(1-p)^n\right\}, \\ Var(\mathcal{C}_k(GF(2)^n),p) &= \frac{(2^k-1)(2^n-2^k)}{(2^n-1)(2^n-2)}\left\{\left(p^2+(1-p)^2\right)^n-(1-p)^{2n}\right\} \\ &\quad -\frac{(2^k-1)(2^n-2^k)}{(2^n-1)^2(2^n-2)}\left\{1-(1-p)^n\right\}^2. \end{aligned}$$

From Corollary 3.2.1 we see that for $\mathcal{C}_k(GF(2)^n)$, the set of all $[n,k]$ codes, and p close to 0 we get

$$\begin{aligned} P_{\text{ue}}(\mathcal{C}_k(GF(2)^n),p) &\approx \frac{2^k-1}{2^n-1}np, \\ Var(\mathcal{C}_k(GF(2)^n),p)) &\approx \frac{(2^k-1)(2^n-2^k)}{(2^n-1)(2^n-2)}\left\{n-\frac{n^2}{2^n-1}\right\}p^2. \end{aligned}$$

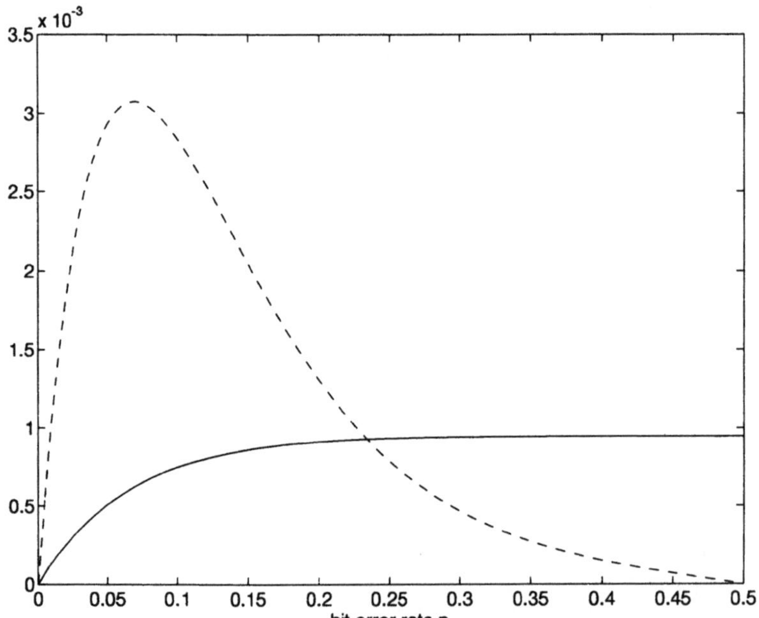

Figure 3.2 Plot of $P_{\text{ue}}(\mathcal{C}_5(GF(2)^{15}),p)$ (in solid) and $\sigma(p)$ (dashed).

In particular, if $n-k$ is large, then

$$\begin{aligned} P_{\text{ue}}(\mathcal{C}_k(GF(2)^n),p) &\approx \frac{n}{2^{n-k}}p, \\ \sigma(p) &\approx \frac{\sqrt{n}}{2^{(n-k)/2}}p, \end{aligned}$$

and so $\sigma(p)$ is large compared to $P_{\text{ue}}(\mathcal{C}_k(GF(2)^n),p)$, that is, there is a large spread in the values of $P_{\text{ue}}(p)$. As an illustration we give in Figure 3.2 a plot of $P_{\text{ue}}(\mathcal{C}_5(GF(2)^{15}),p)$ and the corresponding $\sigma(p)$.

Remark. Corollary 3.2.1 implies in particular that for any given p there exists an $[n,k]$ code C such that

$$P_{\text{ue}}(p) \le \frac{2^k-1}{2^n-1}\left\{1-(1-p)^n\right\}.$$

Next we will consider the mean probability of undetected error for some classes of systematic codes (the variance can be found in a similar way). Let $\mathcal{SYS}(n,k)$ be the set of systematic $[n,k]$ codes and $\mathcal{SYS}(n,k,d)$ the set of systematic $[n,k,d]$ codes.

Lemma 3.2.2 *We have*

$$\begin{aligned} A_i(\mathcal{SYS}(n,k,d)) &= 0 \quad \textit{for } 1 \le i \le d-1, \\ A_i(\mathcal{SYS}(n,k,d)) &\le \frac{1}{2^{n-k} - \sum_{l=0}^{d-2}\binom{n-1}{l}}\left\{\binom{n}{i} - \binom{n-k}{i}\right\}, \end{aligned}$$

with equality for $d=1$.

Proof: Let $i \ge d$. A vector of the form $(\bar{0}|\bar{v})$, where $\bar{v} \in GF(2)^k \setminus \{\bar{0}\}$, is not contained in any systematic code. Hence the number of vectors $\bar{x}$ of weight i which can be contained in a systematic code is $\binom{n}{i} - \binom{n-k}{i}$. Let $\bar{x}$ be such a vector. It remains to show that

$$\alpha(\bar{x}) \le \frac{1}{2^{n-k} - \sum_{l=0}^{d-2}\binom{n-1}{l}}. \tag{3.4}$$

The vector $\bar{x}$ contains at least one 1 in the first k positions. Without loss of generality, we may assume that it is 1 in the first position, i.e. $\bar{x} = (1|\bar{z}|\bar{v})$, where $\bar{z} \in GF(2)^{k-1}$. Let

$$\begin{pmatrix} 1 & \bar{z} & \bar{v} \\ \bar{0}^{\mathrm{t}} & I_{k-1} & P \end{pmatrix}$$

be a generator matrix for a code in $\mathcal{SYS}(n,k,d)$ which contains $\bar{x}$. Then $(I_{k-1}|P)$ generates a code in $\mathcal{SYS}(n-1,k-1,d)$. Hence

$$\alpha(\bar{x}) \leq \frac{\#\mathcal{SYS}(n-1,k-1,d)}{\#\mathcal{SYS}(n,k,d)}. \tag{3.5}$$

On the other hand, if $G = (I_{k-1}|P)$ is the generator matrix for a code in $\mathcal{SYS}(n-1,k-1,d)$, then any combination of $d-1$ or less columns in the corresponding check matrix $(P^{\mathrm{t}}|I_{n-k})$ has a non-zero sum. Let $\bar{y}^{\mathrm{t}}$ be a non-zero vector in $GF(2)^{n-k}$ different from all sums of $d-2$ or less of these columns; $\bar{y}$ can be chosen in at least $2^{n-k} - 1 - \sum_{l=1}^{d-2}\binom{n-1}{l}$ distinct ways. Each choice of $\bar{y}$ gives a check matrix $(\bar{y}^{\mathrm{t}}|P^{\mathrm{t}}|I_{n-k})$ such that any combination of $d-1$ or less columns has a non-zero sum, that is, a check matrix for a code in $\mathcal{SYS}(n,k,d)$. Therefore,

$$\frac{\#\mathcal{SYS}(n,k,d)}{\#\mathcal{SYS}(n-1,k-1,d)} \geq 2^{n-k} - \sum_{l=0}^{d-2}\binom{n-1}{l}. \tag{3.6}$$

Combining (3.5) and (3.6), we get (3.4). For $d = 1$ we get equality both in (3.5) and (3.6).

Theorem 3.2.4 *Let $\mathcal{C} = \mathcal{SYS}(n,k,d)$, the set of all systematic $[n,k,d]$ codes. Then*

$$P_{\mathrm{ue}}(\mathcal{C},p) \leq \frac{1-(1-p)^k - \sum_{i=1}^{d-1}\left\{\binom{n}{i} - \binom{n-k}{i}\right\}p^i(1-p)^{n-i}}{2^{n-k} - \sum_{l=0}^{d-2}\binom{n-1}{l}}$$

with equality for $d = 1$.

Proof: Let

$$\beta = \frac{1}{2^{n-k} - \sum_{l=0}^{d-2}\binom{n-1}{l}}.$$

By Lemma 3.2.2 we get

$$\begin{aligned} P_{\mathrm{ue}}(\mathcal{C},p) &\le \beta\left\{\sum_{i=d}^{n}\binom{n}{i}p^i(1-p)^{n-i}-\sum_{i=d}^{r}\binom{n-k}{i}p^i(1-p)^{n-i}\right\} \\ &= \beta\left\{1-(1-p)^k-\sum_{i=1}^{d-1}\left\{\binom{n}{i}-\binom{n-k}{i}\right\}p^i(1-p)^{n-i}\right\}. \end{aligned}$$

How good is this upper bound? It is exact for $d=1$. We will next derive the exact value of $P_{\mathrm{ue}}(\mathcal{SYS}(n,k,2),p)$ and compare this with the upper bound in Theorem 3.2.4.

Theorem 3.2.5 *Let $\mathcal{C}=\mathcal{SYS}(n,k,2)$. For $1\le i\le n$ we have*

$$A_i(\mathcal{C})=\frac{1}{2^r}\left\{\binom{n}{i}+2^r\binom{k}{i}\zeta^i-\sum_{l=0}^{i}\binom{k}{l}\binom{r}{i-l}\zeta^l\right\},$$

where $r=n-k$ and $\zeta=-1/(2^r-1)$, and

$$P_{\mathrm{ue}}(\mathcal{C},p)=\frac{1}{2^r}\left\{1-\delta(p)^k-2^r(1-p)^n+2^r(1-p)^r\delta(p)^k\right\},$$

where

$$\delta(p)=1-\frac{2^rp}{2^r-1}.$$

Proof: We see that $(I_k|P)$ is a generator matrix for a code in $\mathcal{C}$ if and only if all the rows of P are non-zero. In particular, $\#\mathcal{C}=(2^r-1)^k$. Let $\beta(i)$ be the number of ways $\bar{0}\in GF(2)^r$ can be written as an ordered sum of i vectors from $GF(2)^r\setminus\{\bar{0}\}$ and let $\gamma(i)$ be the number of ways a non-zero $\bar{v}\in GF(2)^r$ can be written as such a sum. A simple induction shows that

$$\beta(i)=\frac{2^r-1}{2^r}\left\{(2^r-1)^{i-1}-(-1)^{i-1}\right\},$$

$$\gamma(i)=\frac{1}{2^r}\left\{(2^r-1)^i-(-1)^i\right\}.$$

The number of codes in $\mathcal{C}$ containing a vector $(\bar{u}|\bar{0})$ where $w_H(\bar{u})=i$ is $\beta(i)(2^r-1)^{k-i}$ and the number of such vectors is $\binom{k}{i}$. For $0<l<k$, the number of codes

containing a vector $(\bar{u}|\bar{v})$ where $w_H(\bar{u}) = l$ and $w_H(\bar{v}) = i-l$ is $\gamma(l)(2^r-1)^{k-i+l}$, and the number of such vectors is $\binom{k}{l}\binom{r}{i-l}$. Hence

$$\begin{aligned} A_i(C) &= \binom{k}{i}\frac{1}{(2^r-1)^i}\cdot\frac{2^r-1}{2^r}\left\{(2^r-1)^{i-1}-(-1)^{i-1}\right\} \\ &\quad +\sum_{l=1}^{i-1}\binom{k}{l}\binom{r}{i-l}\frac{1}{(2^r-1)^{i-l}}\cdot\frac{1}{2^r}\left\{(2^r-1)^{i-l}-(-1)^{i-l}\right\} \\ &= \frac{1}{2^r}\left\{\sum_{l=1}^{i-1}\binom{k}{l}\binom{r}{i-l}-\binom{k}{i}\varsigma^{i-1}-\sum_{l=1}^{i-1}\binom{k}{l}\binom{r}{i-l}\varsigma^l\right\}. \end{aligned}$$

Since

$$\sum_{l=1}^{i-1}\binom{k}{l}\binom{r}{i-l}=\binom{n}{i}-\binom{k}{i}-\binom{r}{i}$$

and

$$\varsigma^{i-1}=-(2^r-1)\varsigma^i,$$

we get the expression for A_i given in the theorem. Putting this into the expression $P_{\mathrm{ue}}(C,p)=\sum_{i=1}^{n}A_i(C)p^i(1-p)^{n-i}$ and rearranging, we get the expression for $P_{\mathrm{ue}}(C,p)$ given in the theorem.

The upper bound in Theorem 3.2.4 is exact for $p=0$ and too large by the quantity $\frac{2^k-k-1}{(2^r-1)2^n}$ for $p=\frac{1}{2}$. The difference between the upper bound and the exact value appears to increase monotonously when p increases from 0 to $\frac{1}{2}$. The exact value of $P_{\mathrm{ue}}(\mathcal{SYS}(n,k,d),p)$ for general d is not known, and it is probably quite complicated both to derive and express.

Theorem 3.2.6 *Let C be the set of all systematic even weight $[n,k]$ codes. Then*

$$\begin{aligned} A_i(C) &= 0 \quad \textit{for odd } i, \\ A_i(C) &= \frac{1}{2^{r-1}}\left\{\binom{n}{i}-\binom{n-k}{i}\right\} \quad \textit{for even } i>0, \end{aligned}$$

and

$$P_{\mathrm{ue}}(C,p)=\frac{1}{2^{n-k}}\left\{1+(1-2p)^n-(1-p)^k-(1-2p)^{n-k}(1-p)^k\right\}.$$

Proof: The proof is similar, and we only give a sketch. All the rows of P have to have odd weight. For a vector $\bar{x} = (\bar{u}|\bar{v})$, where $\bar{u} \neq \bar{0}$ and $w_H(\bar{x})$ is even, we have

$$\alpha(\bar{x}) = 2^{-(r-1)},$$

and for even i there are $\binom{n}{i} - \binom{n-k}{i}$ such vectors of weight i. This proves the expression for $A_i(\mathcal{C})$. Further,

$$\begin{aligned} P_{\text{ue}}(\mathcal{C}, p) &= 2^{-(r-1)} \sum_{i=1}^{\lfloor n/2 \rfloor} \left\{ \binom{n}{2i} - \binom{n-k}{2i} \right\} p^{2i}(1-p)^{n-2i} \\ &= 2^{-r} \left\{ 1 + (1-2p)^n - (1-p)^k - (1-2p)^r(1-p)^k \right\}. \end{aligned}$$

3.3 THE WORST-CASE ERROR PROBABILITY

In many applications we do not know the value of p, at least not exactly. In other applications we may want to use one code for several different values of p. Therefore, we will be interested in finding the worst-case error probability

$$P_{\text{wc}}(C, a, b) = \max_{a \le p \le b} P_{\text{ue}}(C, p)$$

for some interval $[a, b] \subseteq [0, 1]$. First we give an upper bound.

Theorem 3.3.1 *Let C be an $[n, k]$ code. Then*

$$P_{\text{wc}}(C, a, b) \le \sum_{i=1}^{n} A_i \mu_{ni}(a, b) \tag{3.7}$$

where $A_0, A_1, \ldots, A_n$ is the weight distribution of C and

$$\mu_{ni}(a, b) = \begin{cases} a^i(1-a)^{n-i} & \text{if } \frac{i}{n} < a, \\ \left(\frac{i}{n}\right)^i \left(1 - \frac{i}{n}\right)^{n-i} & \text{if } a \le \frac{i}{n} \le b, \\ b^i(1-b)^{n-i} & \text{if } \frac{i}{n} > b. \end{cases}$$

If C is a constant weight code, we have equality in (3.7).

Proof: First we note that the function $p^i(1-p)^{n-i}$ is increasing on the interval $\left[0, \frac{i}{n}\right]$ and decreasing on the interval $\left[\frac{i}{n}, 1\right]$. Hence

$$\max_{a \le p \le b} \left\{p^i(1-p)^{n-i}\right\} = \mu_{ni}(a,b),$$

and so

$$\begin{aligned} P_{\mathrm{wc}}(C,a,b) &= \max_{a \le p \le b} \sum_{i=1}^{n} A_i p^i (1-p)^{n-i} \\ &\le \sum_{i=1}^{n} A_i \max_{a \le p \le b} p^i(1-p)^{n-i} \\ &= \sum_{i=1}^{n} A_i \mu_{ni}(a,b). \end{aligned}$$

If C is a constant weight we get equality.

The bound in Theorem 3.3.1 may be improved by subdividing the interval $[a,b]$: let $a = a_0 < a_1 < \cdots < a_m = b$. Then clearly

$$P_{\mathrm{wc}}(C,a,b) = \max_{1 \le j \le m} P_{\mathrm{wc}}(C, a_{j-1}, a_j).$$

It is easy to see that if

$$F = \sum_{i=1}^{n} A_i \max_{a \le p \le b} \left| \frac{d}{dp}\left(p^i(1-p)^{n-i}\right)\right|,$$

then

$$\begin{aligned} P_{\mathrm{wc}}(C,a_{j-1},a_j) &\le \sum_{i=1}^{n} A_i \mu_{ni}(a_{j-1},a_j) \\ &\le P_{\mathrm{ue}}(C,a_{j-1}) + (a_j - a_{j-1})F \\ &\le P_{\mathrm{wc}}(C,a_{j-1},a_j) + (a_j - a_{j-1})F. \end{aligned}$$

Hence we can get as sharp a bound as we want by subdividing the interval sufficiently.

Remark. It is possible to find an upper bound similar to the bound in Theorem 3.3.1, but in terms of the weight distribution $B_0, B_1, \ldots, B_n$ of the dual code:

$$P_{\mathrm{wc}}(C,a,b) \le \frac{1}{2^{n-k}} \sum_{i=0}^{n} B_i \nu_{ni}(a,b)$$

where $\nu_{ni}(a,b)$ is the maximum of the function $f_{in}(p) = (1-2p)^i - (1-p)^n$ on the interval $[a,b]$. As in the proof above, the maximum is obtained in a, in b or in a maximum of $f_{in}(p)$. In most cases we can not find explicit expressions for the maxima of $f_{in}(p)$, but we can find as good approximations as we like. If we only want to find the worst-case of one code on one interval, it is probably a better idea to do this directly. However, once we have computed the maxima of f_{in} for $i = 0, 1, \ldots, n$, then it is a simple task to compute $\nu_{ni}(a,b)$ for any a and b, and since this is independent of any code, we can compute the worst-case probability of any code for which we know the weight distribution of the dual code (of course, an alternative is to compute the weight distribution of the code itself using MacWilliam's identity and then using Theorem 3.3.1. However, it may be simpler in some cases to use the attack sketched above).

Next we consider a bound on the average value of $P_{\rm wc}(C,a,b)$ over some class $\mathcal{C}$. Let

$$P_{\rm wc}(\mathcal{C},a,b) = \frac{1}{\#\mathcal{C}} \sum_{c\in\mathcal{C}} P_{\rm wc}(C,a,b).$$

From Theorem 3.3.1 we immediately get the following bound.

Theorem 3.3.2 *Let $\mathcal{C}$ be a class of $[n,k]$ codes. Then*

$$P_{\rm wc}(\mathcal{C},a,b) \le \sum_{i=1}^{n} A_i(\mathcal{C})\mu_{ni}(a,b).$$

We will give a result which relates $P_{\rm wc}(\mathcal{C}_k(K),a,b)$ and $P_{\rm wc}(\mathcal{C}_l(K),a,b)$.

Lemma 3.3.1 *For any $[n,k]$ code C and any $l \le k$ we have*

$$P_{\rm ue}(C,p) = \frac{1}{\left[{k-1 \atop l-1}\right]} \sum_{D\in\mathcal{C}_l(C)} P_{\rm ue}(D,p).$$

Proof: We have

$$\begin{aligned} P_{\rm ue}(C,p) &= \sum_{\substack{\bar{x}\in C\\ \bar{x}\neq\{\bar{0}\}}} p^{w_{\rm H}(\bar{x})}(1-p)^{n-w_{\rm H}(\bar{x})} \\ &= \sum_{\substack{\bar{x}\in C\\ \bar{x}\neq\{\bar{0}\}}} p^{w_{\rm H}(\bar{x})}(1-p)^{n-w_{\rm H}(\bar{x})} \frac{1}{\left[{k-1 \atop l-1}\right]} \sum_{\substack{D\in\mathcal{C}_l(C)\\ \bar{x}\in D}} 1 \end{aligned}$$

$$= \frac{1}{\left[{k-1 \atop l-1}\right]} \sum_{D \in \mathcal{C}_l(C)} \sum_{\substack{\bar{x} \in D \\ \bar{x} \neq \{\bar{0}\}}} p^{w_H(\bar{x})}(1-p)^{n-w_H(\bar{x})}$$

$$= \frac{1}{\left[{k-1 \atop l-1}\right]} \sum_{D \in \mathcal{C}_l(C)} P_{ue}(D,p).$$

Lemma 3.3.2 *For any $[n,k]$ code C and any $l \leq k$ we have*

$$P_{wc}(C,a,b) \leq \frac{1}{\left[{k-1 \atop l-1}\right]} \sum_{D \in \mathcal{C}_l(C)} P_{wc}(D,a,b).$$

Proof: By definition and Lemma 3.3.1 we get

$$\begin{aligned} P_{wc}(C,a,b) &= \max_{a \leq p \leq b} P_{ue}(C,p) \\ &= \frac{1}{\left[{k-1 \atop l-1}\right]} \max_{a \leq p \leq b} \left\{ \sum_{D \in \mathcal{C}_l(C)} P_{ue}(D,p) \right\} \\ &\leq \frac{1}{\left[{k-1 \atop l-1}\right]} \sum_{D \in \mathcal{C}_l(C)} \max_{a \leq p \leq b} P_{ue}(D,p) \\ &= \frac{1}{\left[{k-1 \atop l-1}\right]} \sum_{D \in \mathcal{C}_l(C)} P_{wc}(D,a,b). \end{aligned}$$

Theorem 3.3.3 *Let K be an $[n,\kappa]$ code. For $1 \leq l \leq k \leq \kappa$ we have*

$$P_{wc}(\mathcal{C}_k(K),a,b) \leq \frac{2^k-1}{2^l-1} P_{wc}(\mathcal{C}_l(K),a,b).$$

Proof: By definition and Lemma 3.3.2 we get

$$\begin{aligned} P_{wc}(\mathcal{C}_k(K),a,b) &= \frac{1}{\left[{\kappa \atop k}\right]} \sum_{C \in \mathcal{C}_k(K)} P_{wc}(C,a,b) \\ &\leq \frac{1}{\left[{\kappa \atop k}\right]\left[{k-1 \atop l-1}\right]} \sum_{C \in \mathcal{C}_k(K)} \sum_{D \in \mathcal{C}_l(C)} P_{wc}(D,a,b) \\ &= \frac{1}{\left[{\kappa \atop k}\right]\left[{k-1 \atop l-1}\right]} \sum_{D \in \mathcal{C}_l(K)} P_{wc}(D,a,b) \sum_{\substack{C \in \mathcal{C}_k(K) \\ D \subseteq C}} 1 \end{aligned}$$

$$= \frac{1}{\begin{bmatrix}\kappa \\ k\end{bmatrix}\begin{bmatrix}k-1 \\ l-1\end{bmatrix}} \sum_{D \in \mathcal{C}_l(K)} P_{\mathrm{wc}}(D,a,b) \begin{bmatrix}\kappa - l \\ k - l\end{bmatrix}$$

$$= \frac{\begin{bmatrix}\kappa-l \\ k-l\end{bmatrix}\begin{bmatrix}n \\ l\end{bmatrix}}{\begin{bmatrix}\kappa \\ k\end{bmatrix}\begin{bmatrix}k-1 \\ l-1\end{bmatrix}} P_{\mathrm{wc}}(\mathcal{C}_l(K),a,b) = \frac{2^k-1}{2^l-1} P_{\mathrm{wc}}(\mathcal{C}_l(K),a,b).$$

We now consider the worst-case over $[0,1]$. Since

$$\mu_{ni}(0,1) = \left(\frac{i}{n}\right)^i \left(1-\frac{i}{n}\right)^{n-i},$$

we get the following corollary to Theorem 3.3.2.

Corollary 3.3.1 *For a set of $[n,k]$ codes $\mathcal{C}$ we have*

$$P_{\mathrm{wc}}(\mathcal{C},0,1) \le \sum_{i=1}^{n} A_i(\mathcal{C}) \left(\frac{i}{n}\right)^i \left(1-\frac{i}{n}\right)^{n-i}.$$

Some special cases are given in the following corollaries.

Corollary 3.3.2 *We have*

$$P_{\mathrm{wc}}(\mathcal{C}_k(GF(2)^n),0,1) \le \frac{2^k-1}{2^n-1} S_n,$$

where

$$S_n = \sum_{i=1}^{n} \binom{n}{i} \left(\frac{i}{n}\right)^i \left(1-\frac{i}{n}\right)^{n-i}. \tag{3.8}$$

Corollary 3.3.3 *We have*

$$P_{\mathrm{wc}}(\mathcal{SYS}(n,k),0,1) \le \frac{1}{2^{n-k}}(S_n - S_{n,k}),$$

where

$$S_{n,k} = \sum_{i=1}^{n} \binom{n-k}{i} \left(\frac{i}{n}\right)^i \left(1-\frac{i}{n}\right)^{n-i}. \tag{3.9}$$

Some relations for the sums S_n and $S_{n,k}$ are the following:

$$\sqrt{\frac{\pi n}{2}} - \frac{1}{2} < S_n < \sqrt{\frac{\pi n}{2}} - \frac{1}{3} + \frac{0.108918}{\sqrt{n}}, \tag{3.10}$$

$$S_n = \sqrt{\frac{\pi n}{2}} - \frac{1}{3} + \frac{\sqrt{2\pi}}{24\sqrt{n}} + \left(\frac{1}{n}\right) \quad \text{when} \quad n \to \infty, \tag{3.11}$$

$$S_{n,k} = \frac{1}{2^{2k}}\binom{2k}{k}\sqrt{\frac{\pi n}{2}} - \frac{2}{3} + \left(\frac{1}{\sqrt{n}}\right) \quad \text{when} \quad n \to \infty, \tag{3.12}$$

for $k \geq 1$,

$$S_{n,1} = \frac{1}{2} S_n. \tag{3.13}$$

For example,

$$P_{\text{wc}}(\mathcal{C}_1(GF(2)^n), 0, 1) = \frac{1}{2^n - 1} S_n,$$

and

$$P_{\text{wc}}(\mathcal{SYS}(n, 1), 0, 1) = \frac{1}{2^n} S_n.$$

For the worst-case over $[0, 1/2]$ we get

$$\begin{aligned} \mu_{ni}(0, 1/2) &= \left(\frac{i}{n}\right)^i \left(1 - \frac{i}{n}\right)^{n-i} \quad \text{for} \quad i \leq \frac{n}{2}, \\ \mu_{ni}(0, 1/2) &= \frac{1}{2^n} \quad \text{for} \quad i \geq \frac{n}{2}. \end{aligned}$$

Since

$$\sum_{i=1}^{\lceil n/2 \rceil} \left(\frac{i}{n}\right)^i \left(1 - \frac{i}{n}\right)^{n-i} + \sum_{\lceil n/2 \rceil + 1}^{n} \frac{1}{2^n} = \frac{1}{2} S_n$$

we get

$$P_{\text{wc}}(\mathcal{C}_k(GF(2)^n), 0, 1/2) \leq \frac{2^k - 1}{2(2^n - 1)} S_n. \tag{3.14}$$

In particular, for given n and k, there exists an $[n, k]$ code C such that

$$P_{\text{wc}}(C, 0, 1/2) < \frac{2^k - 1}{2(2^n - 1)} \left(\sqrt{\frac{\pi n}{2}} - \frac{1}{3} + \frac{0.108918}{\sqrt{n}}\right).$$

We do believe that the stronger result is true, that there exists an $[n,k]$ code such that

$$P_{\mathrm{wc}}(C,0,1/2) = \frac{2^k-1}{2^n};$$

that is, a good code.

The bound (3.14) shows that an average code is not too bad. However, some codes may be very bad.

Theorem 3.3.4 *For each $N > 0$ and each $\epsilon \in \left(0, \frac{1}{4}\right)$, there exists a code C such that*

$$P_{\mathrm{ue}}(C,p) > N \cdot P_{\mathrm{ue}}\left(C, \frac{1}{2}\right)$$

for all $p \in \left[\epsilon, \frac{1}{2} - \epsilon\right]$.

Proof: Let $C_n = \{(0|\bar{0}), (1|\bar{0})\} \subseteq GF(2)^n$. Then

$$P_{\mathrm{ue}}(C_n,p) = f_n(p) = p(1-p)^{n-1}.$$

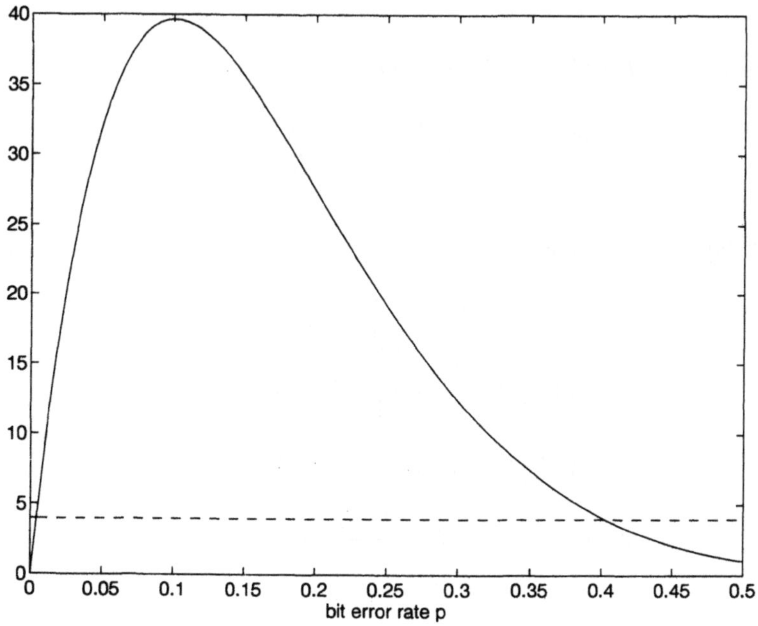

Figure 3.3 Plot of $f_{10}(p)/f_{10}(0.5)$.

Since

$$\frac{d}{dp} f_n(p) = (1-p)^{n-2}(1-np),$$

$f_n(p)$ is increasing on $\left[0, \frac{1}{n}\right]$ and then decreasing. Choose $n \geq \frac{1}{\epsilon}$ such that

$$\frac{f_n\left(\frac{1}{2}-\epsilon\right)}{f_n\left(\frac{1}{2}\right)} = (1-2\epsilon)(1+2\epsilon)^{n-1} > N.$$

Then C_n has the property stated in the theorem.

Example. For $N = 4$ and $\epsilon = 0.1$, we can choose $n = 10$. This is illustrated in Figure 3.3.

3.4 P_{UE} FOR A LINEAR CODE AND ITS DUAL

There is a relation between the the probability of undetected error for a linear code and its dual code, as stated in the next theorem.

Theorem 3.4.1 *Let C be an $[n, k]$ code. Then*

$$P_{\mathrm{ue}}(C,p) = 2^k(1-p)^n P_{\mathrm{ue}}\left(C^{\perp}, \frac{1-2p}{2-2p}\right) + 2^{k-n} - (1-p)^n.$$

Proof: From Theorem 2.3.1 we get

$$\begin{aligned}
P_{\mathrm{ue}}(C,p) &= (1-p)^n\left\{A_C\left(\frac{p}{1-p}\right) - 1\right\} \\
&= (1-p)^n\left\{A_C\left(1-2\frac{1-2p}{2-2p}\right) - 1\right\} \\
&= (1-p)^n\left\{2^k P_{\mathrm{ue}}\left(C^{\perp}, \frac{1-2p}{2-2p}\right) + 2^k\left(1-\frac{1-2p}{2-2p}\right)^n - 1\right\} \\
&= 2^k(1-p)^n P_{\mathrm{ue}}\left(C^{\perp}, \frac{1-2p}{2-2p}\right) + 2^{k-n} - (1-p)^n.
\end{aligned}$$

The dual of a proper code may not be proper, or even good, as shown by the next example.

Example. Let C be the $[15,12]$ code with parity check matrix

$$H = \begin{pmatrix} 11111 & 00000 & 00000 \\ 00000 & 11111 & 00000 \\ 00000 & 00000 & 11111 \end{pmatrix}.$$

Then

$$\begin{aligned} P_{\mathrm{ue}}(C,p) &= \frac{1}{8}\{1+3(1-2p)^5+3(1-2p)^{10}+(1-2p)^{15}\}-(1-p)^{15}, \\ P_{\mathrm{ue}}(C^\perp,p) &= 3p^5(1-p)^{10}+3p^{10}(1-p)^5+p^{15}. \end{aligned}$$

In particular, $C^\perp$ is not good since, for example,

$$P_{\mathrm{ue}}\left(C^\perp,\frac{1}{3}\right)=\frac{3169}{14348907}\approx 0.000221 > P_{\mathrm{ue}}\left(C^\perp,\frac{1}{2}\right)=\frac{7}{32768}\approx 0.000214.$$

On the other hand, C is proper. The graphs of $P_{\mathrm{ue}}\left(C^\perp,p\right)$ and $P_{\mathrm{ue}}\left(C,p\right)$ are shown in Figure 3.4.

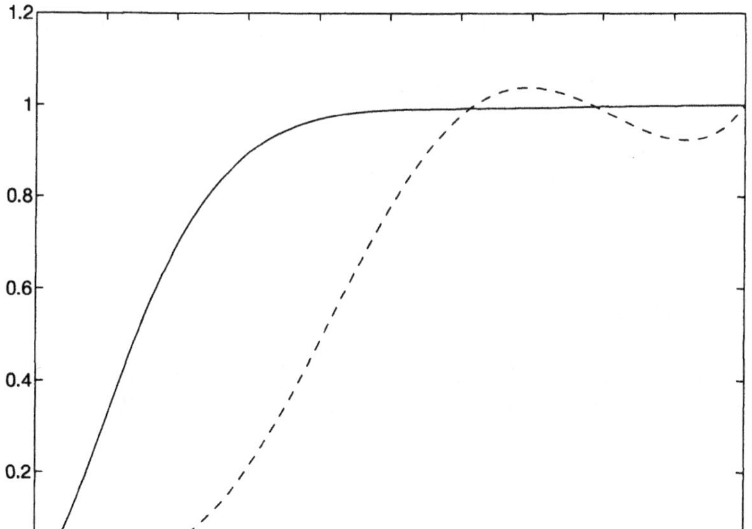

Figure 3.4 Plot of $P_{\mathrm{ue}}\left(C,p\right)/P_{\mathrm{ue}}\left(C,0.5\right)$ (solid line) and $P_{\mathrm{ue}}\left(C^\perp,p\right)/P_{\mathrm{ue}}\left(C^\perp,0.5\right)$ (dashed line).

There are a couple of conditions, however, such that if C satisfies the condition, then so does $C^{\perp}$.

Theorem 3.4.2 *Let C be an $[n,k]$ code.*

1. *If $P_{\mathrm{ue}}(C,p) \le 2^{-(n-k)}$ for all $p \in \left[0,\frac{1}{2}\right]$, then $P_{\mathrm{ue}}(C^{\perp},p) \le 2^{-k}$ for all $p \in \left[0,\frac{1}{2}\right]$.*
2. *If $P_{\mathrm{ue}}(C,p) \le 2^{-(n-k)}\left\{1-(1-p)^k\right\}$ for all $p \in \left[0,\frac{1}{2}\right]$, then $P_{\mathrm{ue}}(C^{\perp},p) \le 2^{-k}\left\{1-(1-p)^{n-k}\right\}$ for all $p \in \left[0,\frac{1}{2}\right]$.*
3. *If $P_{\mathrm{ue}}(C,p) \le \frac{2^k-1}{2^n-1}\{1-(1-p)^n\}$ for all $p \in \left[0,\frac{1}{2}\right]$, then $P_{\mathrm{ue}}(C^{\perp},p) \le \frac{2^{n-k}-1}{2^n-1}\{1-(1-p)^n\}$ for all $p \in \left[0,\frac{1}{2}\right]$.*

Proof: Assume that $P_{\mathrm{ue}}(C,p) \le 2^{-(n-k)}$ for all $p \in \left[0,\frac{1}{2}\right]$. From Theorem 3.4.1 we get

$$\begin{aligned} P_{\mathrm{ue}}(C^{\perp},p) &= 2^{n-k}(1-p)^n P_{\mathrm{ue}}\left(C,\frac{1-2p}{2-2p}\right) + 2^{-k} - (1-p)^n \\ &\le 2^{n-k}(1-p)^n 2^{-(n-k)} + 2^{-k} - (1-p)^n \\ &= 2^{-k}. \end{aligned}$$

The proof of the other relations are similar.

3.5 GENERAL BOUNDS

In many cases one is not able to find the weight distribution. Therefore, we give a number of lower and upper bounds on $P_{\mathrm{ue}}[n,k,p]$ and $P_{\mathrm{ue}}(n,M,p)$ which may be used to estimate P_{ue}.

Lower Bounds

Theorem 3.5.1 *For any n, k, and any $p \in \left[0,\frac{1}{2}\right]$ we have*

$$P_{\mathrm{ue}}[n,k,p] \ge \left(2^k-1\right) p^{\frac{n2^{k-1}}{2^k-1}} (1-p)^{\frac{n\left(2^{k-1}-1\right)}{2^k-1}}.$$

Proof: Let $C = \{\bar{x}_0 = \bar{0}, \bar{x}_1, \cdots, \bar{x}_{2^k-1}\}$ be an $[n,k]$ code and let its support weight be m. Let $t_i = w_H(\bar{x}_i)$, the Hamming weight of the $\bar{x}_i$. Then

$$P_{\mathrm{ue}}(C,p) = \sum_{i=1}^{2^k-1} p^{t_i}(1-p)^{n-t_i}.$$

If j is in the support of C, then half the code words of C has a 1 in position j, the other half has a 0. Hence

$$\sum_{i=1}^{2^k-1} t_i = m2^{k-1}.$$

Let $\lambda_i = p^{t_i}(1-p)^{n-t_i}$. Then $P_{\mathrm{ue}}(C,p) = \sum_{i=1}^{2^k-1} \lambda_i$. Moreover

$$\prod_{i=1}^{2^k-1} \lambda_i = p^{\sum_{i=1}^{2^k-1} t_i}(1-p)^{(2^k-1)n-\sum_{i=1}^{2^k-1} t_i} = p^{m2^{k-1}}(1-p)^{(2^k-1)n-m2^{k-1}}.$$

If $y_1, y_2, \cdots, y_N$ are real numbers such that $\prod_{i=1}^N y_i = c > 0$, then it is easily seen that $\sum_{i=1}^N y_i \geq Nc^{\frac{1}{N}}$, that is, the minimum is obtained when all the y_i are equal. In particular, we get

$$\begin{aligned} P_{\mathrm{ue}}(C,p) &\geq (2^k-1)p^{m\frac{2^{k-1}}{2^k-1}}(1-p)^{n-m\frac{2^{k-1}}{2^k-1}} \\ &= (2^k-1)(1-p)^n\left(\frac{p}{1-p}\right)^{m\frac{2^{k-1}}{2^k-1}} \\ &\geq (2^k-1)(1-p)^n\left(\frac{p}{1-p}\right)^{n\frac{2^{k-1}}{2^k-1}}. \end{aligned}$$

for all $p \in \left[0, \frac{1}{2}\right]$.

Theorem 3.5.2 *For any n, M, and any $p \in \left[0, \frac{1}{2}\right]$ we have*

$$P_{\mathrm{ue}}(n,M,p) \geq \frac{M}{2^n} - (1-p)^n.$$

Proof: Let C be an (n,M) code. Then

$$P_{\mathrm{ue}}(C,p) = \frac{M}{2^n}A_C^{MW}(1-2p) - (1-p)^n \geq \frac{M}{2^n} - (1-p)^n.$$

Remark. The bound in Theorem 3.5.2 is negative and hence not interesting for $p < 1 - \frac{1}{2}M^{1/n}$. However, for $p \to \frac{1}{2}$ it is asymptotically tight.

For linear codes there is a sharper bound of the same form which follows directly from Corollary 2.2.4.

Theorem 3.5.3 *For any n, k, and any $p \in \left[0, \frac{1}{2}\right]$ we have*

$$P_{\mathrm{ue}}[n,k,p] \geq \frac{1}{2^{n-k}} \prod_{j=k+1}^{n} \left(1 + (1-2p)^j\right) - (1-p)^n.$$

There are a couple of bounds that improve these bounds.

Lemma 3.5.1 *Let C be an (n, M) code, $0 \leq u \leq 1$, and $p \in \left[0, \frac{1}{2}\right]$. Then*

$$P_{\mathrm{ue}}(C,p) \geq (M-1)^{1-\frac{1}{u}} \left\{p^u + (1-p)^u\right\}^{\frac{n}{u}} P_{\mathrm{ue}}\left(C, \frac{p^u}{p^u + (1-p)^u}\right)^{\frac{1}{u}}.$$

Proof: Let

$$p_u = \frac{p^u}{p^u + (1-p)^u}.$$

Then

$$\frac{p_u}{1-p_u} = \left(\frac{p}{1-p}\right)^u.$$

Let $\sum_{i=0}^{n} A_i z^i$ be the distance distribution function of C. Using the Hölder inequality (see e.g. [21, page 62]) we get

$$\begin{aligned}
P_{\mathrm{ue}}(C,p_u) &= (1-p_u)^n \sum_{i=1}^{n} A_i \left(\frac{p}{1-p}\right)^{ui} \\
&= \frac{(1-p)^{un}}{\{p^u + (1-p)^u\}^n} \sum_{i=1}^{n} \left\{A_i \left(\frac{p}{1-p}\right)^i\right\}^u A_i^{1-u} \\
&\leq \frac{(1-p)^{un}}{\{p^u + (1-p)^u\}^n} \left\{\sum_{i=1}^{n} A_i \left(\frac{p}{1-p}\right)^i\right\}^u \left\{\sum_{i=1}^{n} A_i\right\}^{1-u} \\
&= \frac{1}{\{p^u + (1-p)^u\}^n} P_{\mathrm{ue}}(C,p)^u (M-1)^{1-u}.
\end{aligned}$$

Solving for $P_{\mathrm{ue}}(C,p)$, the theorem follows.

Theorem 3.5.4 *Let* $M = \lceil 2^{Rn} \rceil$ *and*

$$\alpha(n,R) = -\frac{R}{\log_2(2^{1-R+\frac{1}{n}} - 1)}.$$

Then

$$P_{\mathrm{ue}}(n,M,p) \geq (M-1)p^{n\alpha(n,R)}(1-p)^{n(1-\alpha(n,R))}.$$

Proof: Let

$$u = \frac{\log_2(2^{1-R+\frac{1}{n}} - 1)}{\log_2\left(\frac{p}{1-p}\right)} = -\frac{R}{\alpha(n,R)\log_2(\frac{p}{1-p})}.$$

Then

$$\frac{M}{2^n} \geq 2^{Rn-n} = 2\left(\frac{(1-p)^u}{p^u + (1-p)^u}\right)^n.$$

Combining this with Lemma 3.5.1 and Theorem 3.5.2 we get

$$\begin{aligned}
P_{\mathrm{ue}}(C,p) &\geq (M-1)^{1-\frac{1}{u}}\{p^u + (1-p)^u\}^{\frac{n}{u}}\left\{\frac{M}{2^n} - \left(\frac{(1-p)^u}{p^u + (1-p)^u}\right)^n\right\}^{\frac{1}{u}} \\
&\geq (M-1)^{1-\frac{1}{u}}(1-p)^n \\
&\geq (M-1)(1-p)^n 2^{-\frac{Rn}{u}} \\
&= (M-1)(1-p)^n\left(\frac{p}{1-p}\right)^{n\alpha(R,n)}.
\end{aligned}$$

A trivial lower bound of a different form for linear codes is the following.

Theorem 3.5.5 *For any* $[n,k,d]$ *code* C *and any* $p \in \left[0, \frac{1}{2}\right]$ *we have*

$$P_{\mathrm{ue}}(C,p) \geq p^d(1-p)^{n-d}.$$

There are a couple of bounds that follows from the fact that

$$p^i(1-p)^{n-i} \geq p^j(1-p)^{n-j} \text{ for } i \leq j \text{ and } p \in \left[0, \frac{1}{2}\right]. \tag{3.15}$$

Theorem 3.5.6 *Let n, $2 \le K \le M$ and $p \in \left[0, \frac{1}{2}\right]$ be given. Then we have*

$$P_{\mathrm{ue}}(n, M, p) \ge \frac{M - K + 1}{K - 1} p^{d(n,K)} (1 - p)^{n-d(n,K)}.$$

Proof: Let C be an (n, M) code and let

$$E = \{(\bar{x}, \bar{y}) \in C^2 \mid 0 < d_H(\bar{x}, \bar{y}) \le d(n, K)\}.$$

We consider C the vertex set and E the edge set of a graph. Let $F \subseteq C$ be an independent set (no two vertices in F are connected by an edge). Then

$$d_H(\bar{x}, \bar{y}) > d(n, K)$$

for all $\bar{x}, \bar{y} \in F$. By the definition of $d(n, K)$, this implies that $\#F \le K - 1$. A result by Turan (see: [14, Theorem 13.4.1]) implies that

$$\#E \ge \frac{(M - K + 1)M}{K - 1}.$$

Hence

$$\begin{aligned} P_{\mathrm{ue}}(C, p) &\ge \frac{1}{M} \sum_{(\bar{x},\bar{y}) \in E} p^{d_H(\bar{x},\bar{y})} (1 - p)^{n - d_H(\bar{x},\bar{y})} \\ &\ge \frac{M - K + 1}{K - 1} p^{d(n,K)} (1 - p)^{n-d(n,K)}. \end{aligned}$$

From Theorems 3.5.6 and 3.5.4 we get the following bounds.

Theorem 3.5.7 *(i) For $0 \le R_1 \le R_2 \le 1$ we have*

$$\overline{\pi}_{\mathrm{ue}}(R_2, p) \le -\delta(R_1) \log_2(p) - (1 - \delta(R_1)) \log_2(1 - p) + R_1 - R_2.$$

For $0 \le R \le 1$ we have

$$\overline{\pi}_{\mathrm{ue}}(R, p) \le \frac{R}{\log_2(2^{1-R} - 1)} \log_2(p) - \left\{1 + \frac{R}{\log_2(2^{1-R} - 1)}\right\} \log_2(1 - p) - R.$$

Upper Bounds

We now turn to upper bounds. From (3.15) we immediately get the next theorem.

Theorem 3.5.8 *For any $[n,k,d]$ code C and any $p \in [0, \frac{1}{2}]$ we have*

$$P_{\mathrm{ue}}(C,p) \le (2^k - 1)\, p^d (1-p)^{n-d}.$$

The bound can be sharpened using upper bounds on A_i. The proof is similar.

Theorem 3.5.9 *Let C be an $[n,k,d]$ code and $p \in [0, \frac{1}{2}]$. Suppose that $A_i(C) \le \alpha_i$ for $d \le i \le n$. Let l be minimal such that*

$$\sum_{i=d}^{l} \alpha_i \ge (2^k - 1)\,.$$

Then

$$P_{\mathrm{ue}}(C,p) \le \sum_{i=d}^{l-1} \alpha_i p^i (1-p)^{n-i} + \left(2^k - 1 - \sum_{i=d}^{l-1} \alpha_i\right) p^l (1-p)^{n-l}.$$

Example. Consider an $[8,4,3]$ code. By Theorem 3.5.8 we get

$$P_{\mathrm{ue}}(C,p) \le 15p^3(1-p)^5.$$

Using the bounds on A_i given in Theorem 2.2.11 we get

$$A_3 \le \left\lfloor \frac{28}{3} \right\rfloor = 9, \quad A_4 \le 14$$

and so

$$P_{\mathrm{ue}}(C,p) \le 9p^3(1-p)^5 + 6p^4(1-p)^4.$$

Theorem 3.5.10 *Let C be an $[n,k,d]$ code and $p \in [0, \frac{1}{2}]$. Then*

$$P_{\mathrm{ue}}(C,p) \le p^d(1-p)^{n-d} \sum_{i=1}^{d} \binom{k}{i} + \sum_{i=d+1}^{k} \binom{k}{i} p^i (1-p)^{n-i}.$$

In particular

$$P_{\mathrm{ue}}(C,p) \le (1-p)^{n-k} - (1-p)^n$$

for any $[n,k]$ code C.

Proof: Let G be a generator matrix for C. Since equivalent codes have the same probability of undetected error, we may assume without loss of generality that $G = (I_k|Q)$ where Q is some $k \times (n-k)$ matrix. We have

$$w_H(\bar{x}G) = w_H(\bar{x}) + w_H(\bar{x}Q) \geq w_H(\bar{x}).$$

Also $w_H(\bar{x}G) \geq d$. Hence

$$\begin{aligned} P_{\text{ue}}(C,p) &= \sum_{\bar{x}\in GF(2)^k\setminus\{\bar{0}\}} p^{w_H(\bar{x}G)}(1-p)^{n-w_H(\bar{x}G)} \\ &\leq \sum_{\substack{\bar{x}\in GF(2)^k \\ 1\leq w_H(\bar{x})\leq d}} p^d(1-p)^{n-d} + \sum_{\substack{\bar{x}\in GF(2)^k \\ d+1\leq w_H(\bar{x})\leq k}} p^{w_H(\bar{x})}(1-p)^{n-w_H(\bar{x})}. \end{aligned}$$

In particular

$$\begin{aligned} P_{\text{ue}}(C,p) &\leq (1-p)^{n-k} \sum_{\bar{x}\in GF(2)^k\setminus\{\bar{0}\}} p^{w_H(\bar{x})}(1-p)^{k-w_H(\bar{x})} \\ &= (1-p)^{n-k}\left(1-(1-p)^k\right). \end{aligned}$$

Remark. If C is the code generated by $\left(I_k|0_{k\times(n-k)}\right)$, where $0_{k\times(n-k)}$ is the $k \times (n-k)$ matrix with all entries zero, then

$$P_{\text{ue}}(C,p) = (1-p)^{n-k} - (1-p)^n.$$

Hence, Theorem 3.5.10 is best possible for $[n,k,1]$ codes.

Examples. If we use Theorem 3.5.10 we get

$$P_{\text{ue}}(C,p) \leq 14p^3(1-p)^5 + p^4(1-p)^4$$

for an $[8,4,3]$ code, and this is weaker then the bound obtained from Theorems 2.2.11 and 3.5.9. In other cases, Theorem 3.5.10 may give a better bound. For example, for a $[14,5,3]$ code C, Theorem 3.5.10 gives

$$P_{\text{ue}}(C,p) \leq 25p^3(1-p)^{11} + 5p^4(1-p)^{10} + p^5(1-p)^9,$$

whereas Theorems 2.2.11 and 3.5.9 give only

$$P_{\text{ue}}(C,p) \leq 30p^3(1-p)^{11} + p^4(1-p)^{10}.$$

The mean results clearly gives upper bounds on $P_{\text{ue}}[n,k,p]$.

Theorem 3.5.11 *If K is any $[n, \kappa]$ code, then*

$$P_{\mathrm{ue}}[n, k, p] \le \frac{2^k - 1}{2^\kappa - 1} P_{\mathrm{ue}}(K, p)$$

for all $p \in \left[0, \frac{1}{2}\right]$. In particular, for all $n \ge \kappa \ge k \ge 1$ we have

$$P_{\mathrm{ue}}[n, k, p] \le \frac{2^k - 1}{2^\kappa - 1} P_{\mathrm{ue}}[n, \kappa, p].$$

Further, for all $n \ge k \ge 1$ we have

$$P_{\mathrm{ue}}[n, k, p] \le \frac{2^k - 1}{2^n - 1} \left\{1 - (1-p)^n\right\},$$

and for all $n > k \ge 1$ we have

$$P_{\mathrm{ue}}[n, k, p] \le \frac{2^k - 1}{2^n - 2} \left\{1 + (1-2p)^n - 2(1-p)^n\right\}.$$

Proof: The main result follows from Theorem 3.2.3, and the special cases by choosing K optimal, $K = GF(2)^n$, and K the $[n, n-1]$ even-weight codes respectively.

Remark. A similar bound for codes of minimum distance at least d can be obtained from 3.2.2.

There are sharper bounds in some cases.

Theorem 3.5.12 *For any integers $n \ge k \ge 1$, any $p \in \left[0, \frac{1}{2}\right]$, and any $u \in (0, 1]$ we have*

$$P_{\mathrm{ue}}[n, k, p] \le \left(\frac{2^k - 1}{2^n - 1} \left\{(p^u + (1-p)^u)^n - (1-p)^{un}\right\}\right)^{\frac{1}{u}}.$$

Proof: Let C be an $[n, k]$ code. Using the natural isomorphism between $GF(2)^n$ and $GF(2^n)$ we may consider C as a subset of $GF(2^n)$. For any non-zero $g \in GF(2^n)$, the set

$$gC = \{gx \mid x \in C\}$$

is again an $[n, k]$ code. For convenience we write

$$J = \left(\frac{2^k - 1}{2^n - 1} \left\{(p^u + (1-p)^u)^n - (1-p)^{un}\right\}\right)^{\frac{1}{u}},$$

the right hand expression in the theorem. Define $\zeta(g)$ by

$$\begin{array}{ll} \zeta(g) = 1 & \text{if } P_{\text{ue}}(gC,p) > J, \\ \zeta(g) = 0 & \text{if } P_{\text{ue}}(gC,p) \le J. \end{array}$$

Note that $\zeta(g) < \frac{P_{\text{ue}}(gC,p)}{J}$ for all g. We will to show that $\zeta(g) = 0$ for at least one g. First we show that

$$\zeta(g) < \sum_{x \in C\setminus\{0\}} \left(\frac{p^{w_H(gx)}(1-p)^{n-w_H(gx)}}{J} \right)^u . \tag{3.16}$$

If $p^{w_H(gx)}(1-p)^{n-w_H(gx)} > J$ for some $x \neq 0$, then (3.16) is clearly satisfied. On the other hand, if $p^{w_H(gx)}(1-p)^{n-w_H(gx)} \le J$ for all $x \neq 0$, then

$$\begin{aligned} \zeta(g) < \frac{P_{\text{ue}}(gC,p)}{J} &= \sum_{x\neq 0} \left(\frac{p^{w_H(gx)}(1-p)^{n-w_H(gx)}}{J} \right) \\ &\le \sum_{x\neq 0} \left(\frac{p^{w_H(gx)}(1-p)^{n-w_H(gx)}}{J} \right)^u , \end{aligned}$$

and (3.16) is again satisfied.

If $x \neq 0$, then gx runs through $GF(2^n)$ when g does. Hence

$$\begin{aligned} \sum_{g\in GF(2^n)\setminus\{0\}} \zeta(g) &< \sum_{g\neq 0}\sum_{x\neq 0} \frac{1}{J^u} p^{uw_H(gx)}(1-p)^{u(n-w_H(gx))} \\ &= \frac{1}{J^u}\sum_{x\neq 0}\sum_{g\neq 0} p^{uw_H(gx)}(1-p)^{u(n-w_H(gx))} \\ &= \frac{1}{J^u}\sum_{x\neq 0} \{(p^u+(1-p)^u)^n-(1-p)^{un}\} \\ &= \frac{2^k-1}{J^u}\{(p^u+(1-p)^u)^n-(1-p)^{un}\} \\ &= 2^n-1. \end{aligned}$$

Therefore $\zeta(g) = 0$ for at least one $g \neq 0$.

Lemma 3.5.2 *If* $R = \frac{k}{n}$ *and*

$$u = \frac{\log_2 \frac{1-\rho(R)}{\rho(R)}}{\log_2 \frac{1-p}{p}},$$

then

$$\left(\frac{2^k-1}{2^n-1}\left\{(p^u+(1-p)^u)^n-(1-p)^{un}\right\}\right)^{\frac{1}{u}} \le p^{n\rho(R)}(1-p)^{n-n\rho(R)}.$$

Proof: For this value of u we have

$$\frac{1-\rho(R)}{\rho(R)} = \left(\frac{1-p}{p}\right)^u$$

and so $p^u+(1-p)^u = \frac{p^u}{\rho(R)}$. Further,

$$2^{R-1} = 2^{-H_2(\rho(R))} = \rho(R)^{\rho(R)}(1-\rho(R))^{1-\rho(R)}.$$

Hence

$$\begin{aligned}
J &\le \left\{2^{k-n}\left(p^u+(1-p)^u\right)^n\right\}^{\frac{1}{u}} \\
&= \left\{\rho(R)^{\rho(R)}(1-\rho(R))^{1-\rho(R)}\frac{p^u}{\rho(R)}\right\}^{\frac{n}{u}} \\
&= \left\{p^u\left(\frac{1-\rho(R)}{\rho(R)}\right)^{1-\rho(R)}\right\}^{\frac{n}{u}} \\
&= \left\{p^u\left(\frac{1-p}{p}\right)^{u(1-\rho(R))}\right\}^{\frac{n}{u}} \\
&= \left\{p^{\rho(R)}(1-p)^{1-\rho(R)}\right\}^n.
\end{aligned}$$

Combining Theorem 3.5.12 and Lemma 3.5.2 we get the following theorem.

Theorem 3.5.13 *If* $p \in \left[0, \frac{1}{2}\right]$ *and* $R = \frac{k}{n} \le C(p) = 1 - H_2(p)$ *(that is,* $\rho(R) \ge p$*), then*

$$P_{\text{ue}}[n,k,p] \le p^{n\rho(R)}(1-p)^{n-n\rho(R)}.$$

Theorem 3.5.14 *For* $0 \le R \le C(p)$ *we have*

$$\underline{\pi}_{\text{ue}}(R,p) \ge -\rho(R)\log_2(p) - (1-\rho(R))\log_2(1-p).$$

For $0 \le R \le 1$ *we have*

$$\underline{\pi}_{\text{ue}}(R,p) \ge 1 - R.$$

Proof: The first bound follows directly from Theorem 3.5.13. From Theorem 3.5.2 we get

$$P_{ue}(n, 2^k, p) \leq P_{ue}[n, k, p] \leq 2^{k-n}$$

and so

$$\frac{\log_2 P_{ue}(n, \lceil 2^{Rn} \rceil, p)}{n} \leq R - 1$$

for $0 \leq R \leq 1$. This proves the second bound. The two bounds coincide when $R = C(p)$; the first bound is stronger when $R < C(p)$.

In Figure 3.5 we give a plot of the lower bound on $\underline{\pi}_{ue}(R, p)$ given in Theorem 3.5.14.

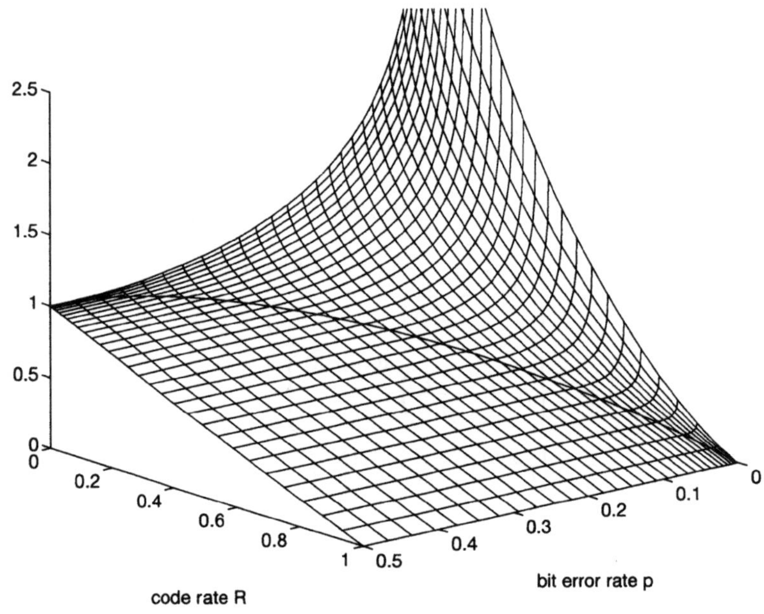

Figure 3.5 Graph of the lower bound on $\underline{\pi}_{ue}(R, p)$.

Theorem 3.5.15 *For* $\log_2(2(1-p)) < R \leq 1$ *we have*

$$\underline{\pi}_{ue}(R, p) = 1 - R.$$

Proof: By Theorem 3.5.2 we have

$$P_{ue}(n, \lceil 2^{Rn} \rceil, p) \geq \frac{\lceil 2^{Rn} \rceil}{2^n} - (1-p)^n \geq 2^{Rn-n} - (1-p)^n.$$

If $R > \log_2(2(1-p))$ we get

$$\lim_{n\to\infty} \frac{(1-p)^n}{2^{Rn-n}} = 0,$$

and so

$$\pi_{\text{ue}}(n, R, p) \leq 1 - R + o\left(\frac{1}{n}\right).$$

Combining this with Theorem 3.5.14 the result follows.

By the asymptotic Gilbert-Varshamov bound, $\delta(R) \geq \rho(R)$ for all R. It is conjectured that $\delta(R) = \rho(R)$ for $0 \leq R \leq C(p)$. If this is the case, then the first of the upper bounds of Theorem 3.5.7 (with $R = R_1 = R_2$) and the lower bound of Theorem 3.5.14 coincide in this range. Hence we have the following theorem.

Theorem 3.5.16 *If $\delta(R) = \rho(R)$ for $0 \leq R \leq C(p)$, then*

$$\pi_{\text{ue}}(R,p) = \begin{cases} -\rho(R)\log_2(p) - (1-\rho(R))\log_2(1-p) & \text{for } 0 \leq R \leq C(p), \\ 1 - R & \text{for } C(p) \leq R \leq 1. \end{cases}$$

Any upper bound on $\delta(R)$ will give an upper bound on $\overline{\pi}_{\text{ue}}(R,p)$. The best upper bounds on $\delta(R)$ known is the LP-bound (see e.g. MacWilliams & Sloane [13, Chapter 17]). Using any upper bound on $\delta(R)$, the first of the bounds of Theorem 3.5.7 is the better for small values of p, the other for larger values of p.

3.6 OPTIMAL CODES

The Dual of an Optimal Code

Theorem 3.6.1 *If C is an $[n,k]$ code which is optimal for p, then $C^\perp$ is optimal for $\frac{1-2p}{2-2p}$. Moreover,*

$$P_{\text{ue}}[n,k,p] = 2^k(1-p)^n P_{\text{ue}}\left[n, n-k, \frac{1-2p}{2-2p}\right] + 2^{k-n} - (1-p)^n.$$

Proof: Theorem 3.4.1 implies that if $P_{\mathrm{ue}}(C_1,p) \le P_{\mathrm{ue}}(C_2,p)$, then

$$P_{\mathrm{ue}}\left(C_1^{\perp}, \frac{1-2p}{2-2p}\right) \le P_{\mathrm{ue}}\left(C_2^{\perp}, \frac{1-2p}{2-2p}\right),$$

and the theorem follows.

Since $\frac{1-2p}{2-2p}$ runs through $\left[0, \frac{1}{2}\right]$ when p does, we have the following corollary.

Corollary 3.6.1 *If C is a code which is optimal for all $p \in \left[0, \frac{1}{2}\right]$, then so is $C^{\perp}$.*

Optimal Codes for Small Dimensions

It is an open question for which n and k there exist $[n,k]$ codes which are optimal for all $p \in \left[0, \frac{1}{2}\right]$. However, for $k \le 4$ such codes do exist for all n; by Corollary 3.6.1 the same is true for $k \ge n-4$. Before we can prove this, we first must show a couple of lemmas. The proof is quite long and we divide it into some lemmas. The proof will be based on Corollary 2.2.1. Let C be an $[n,k]$ code generated by G whose column count function is m. Then

$$P_{\mathrm{ue}}(C,p) = (1-p)^n \sum_{U \in \mathcal{S}_{k,k-1}} \left(\frac{p}{1-p}\right)^{s(U^c,m)}. \tag{3.17}$$

Lemma 3.6.1 *If C is optimal for $p \in \left(0, \frac{1}{2}\right)$, then $m(\bar{0}) = 0$.*

Proof: Suppose that $m(\bar{0}) > 0$. Let $\bar{y}$ be an arbitrary vector in $GF(2)^k \setminus \{\bar{0}\}$. Define m' by

$$\begin{aligned} m'(\bar{0}) &= 0, \\ m'(\bar{y}) &= m(\bar{y}) + m(\bar{0}), \\ m'(\bar{x}) &= m(\bar{x}) \qquad \text{for } \bar{x} \in GF(2)^k \setminus \{\bar{0}, \bar{y}\}, \end{aligned}$$

and let C' be a corresponding code. Then

$$s(U^c, m') = \begin{cases} s(U^c,m) + m(\bar{0})) & \text{if } \bar{y} \in U^c, \\ s(U^c,m) & \text{if } \bar{y} \notin U^c. \end{cases}$$

Hence

$$P_{\text{ue}}(C,p) - P_{\text{ue}}(C',p) = (1-p)^n \left\{1 - \left(\frac{p}{1-p}\right)^{m(\bar{0})}\right\} \sum_U \left(\frac{p}{1-p}\right)^{s(U^c,m)} > 0,$$

where the last sum is over all U such that $\bar{y} \in U \in \mathcal{S}_{k,k-1}$.

Let T be an invertible linear transformation on $GF(2)^k$, and let C_T be a code corresponding to $m \circ T$, that is, if m_T corresponds to C_T, then $m_T(\bar{x}) = m(T(\bar{x}))$.

Lemma 3.6.2 *We have $P_{\text{ue}}(C_T,p) = P_{\text{ue}}(C,p)$ for all p.*

Proof: We have

$$s(U, m_T) = \sum_{\bar{x}\in U} m(T(\bar{x})) = \sum_{T(\bar{x})\in TU} m(T(\bar{x})) = s(TU, m).$$

Since TU runs through $\mathcal{S}_{k,k-1}$ when U does, we get

$$A(C_T, z) = 1 + \sum_{U\in\mathcal{S}_{k,k-1}} z^{n-s(U,m_T)} = 1 + \sum_{U\in\mathcal{S}_{k,k-1}} z^{n-s(U,m)} = A(C,z).$$

Lemma 3.6.3 *If C is an optimal $[n,k]$ code for $p \in \left(0, \frac{1}{2}\right)$ and $k \le 4$, then*

$$|m(\bar{x}) - m(\bar{y})| \le 1 \text{ for } \bar{x}, \bar{y} \in GF(2)^k \setminus \{\bar{0}\}.$$

Proof: We give the proof for $k = 3$. For $k = 1, 2$ the proof is simpler, for $k = 4$ more complicated, but similar. For the details of the proof for $k = 4$ we refer to [103].

To simplify the notation, we write

$$m_1 = m(100), m_2 = m(010), \cdots, m_7 = m(111).$$

First we note that there exists a linear transformation on $GF(2)^3$ such that

$$m_T(100) \le m_T(\bar{x}) \le m_T(010) \text{ for all } \bar{x} \in GF(2)^3 \setminus \{\bar{0}\},$$

$$\text{and } m_T(101) \le m_T(011).$$

Hence by Lemma 3.6.2, we may assume without loss of generality that

$$m_1 \le m_i \le m_2 \text{ for } 1 \le i \le 7,$$

$$m_5 \le m_6.$$

Note that this implies that $m_2 - m_1 \ge m_6 - m_5$. By Corollary 2.2.1 we get

$$\begin{aligned} A_C(z) &= 1 + z^{m_1+m_2+m_5+m_6} + z^{m_1+m_2+m_4+m_7} + z^{m_5+m_6+m_4+m_7} \\ &\quad + z^{m_1+m_5+m_3+m_7} + z^{m_2+m_6+m_3+m_7} \\ &\quad + z^{m_1+m_6+m_3+m_4} + z^{m_2+m_5+m_3+m_4}. \end{aligned}$$

Suppose that the lemma is not true, that is, C is optimal, but $m_2 - m_1 \ge 2$. By Lemma 3.6.1, $m_0 = 0$.

Case I, $m_2 - m_1 = m_6 - m_5$. Define $\tilde{m}$ by

$$\tilde{m}_1 = m_1 + 1,\ \tilde{m}_2 = m_2 - 1,\ \tilde{m}_5 = m_5 + 1,\ \tilde{m}_6 = m_6 - 1,$$

and $\tilde{m}_i = m_i$ for $i \in \{0, 3, 4, 7\}$. Then $\tilde{m}_i \ge 0$ for all i. Let $\tilde{C}$ be a corresponding code. Then

$$A_C(z) - A_{\tilde{C}}(z) = z^{m_1+m_3+m_5+m_7}\left(1 - z^2\right)\left(1 - z^{m_2-m_1+m_6-m_5-2}\right) > 0$$

for $z \in (0, 1)$. Hence $P_{\text{ue}}(C, p) > P_{\text{ue}}(\tilde{C}, p)$, contradicting the optimality of C.

Case II, $m_2 - m_1 > m_6 - m_5$. Define $\tilde{m}$ by

$$\tilde{m}_1 = m_1 + 1,\ \tilde{m}_2 = m_2 - 1,$$

and $\tilde{m}_i = m_i$ for $i \in \{0, 3, 4, 5, 6, 7\}$. Again $\tilde{m}_i \ge 0$ for all i. Let $\tilde{C}$ be a corresponding code. Then

$$\begin{aligned} A_C(z) - A_{\tilde{C}}(z) &= z^{m_1+m_3+m_5+m_7}(1 - z)\left(1 - z^{m_2-m_1-1+m_6-m_5}\right) \\ &\quad + z^{m_1+m_3+m_4+m_6}(1 - z)\left(1 - z^{(m_2-m_1)-(m_6-m_5)-1}\right) \\ &> 0 \end{aligned}$$

for $z \in (0, 1)$. Again $P_{\text{ue}}(C, p) > P_{\text{ue}}(\tilde{C}, p)$, contradicting the optimality of C.

For an $[n, k]$ code C for which $m(\bar{x}) > 0$ for all $\bar{x} \ne \bar{0}$, define C^- to be an $[n - (2^k - 1), k]$ code corresponding to m^- defined by

$$\begin{aligned} m^-(\bar{0}) &= 0, \\ m^-(\bar{x}) &= m(\bar{x}) - 1 \text{ for } \bar{x} \ne \bar{0}. \end{aligned}$$

Then $(C^-)^* = C$ and

$$p^{2^{k-1}}(1-p)^{2^{k-1}-1}P_{\text{ue}}(C^-,p) = P_{\text{ue}}(C,p).$$

Lemma 3.6.4 *Let C be an optimal $[n,k]$ code for $p \in \left(0,\frac{1}{2}\right)$ where $k \leq 4$. Then*
(i) C^* is optimal for p,
(ii) if $n \geq 2^k - 1 + k$, then C^- is defined and it is optimal for p.

Proof: Let C_1 be an optimal $[n + 2^k - 1, k]$ code for p. By Lemma 3.6.3 C_1^- is defined. Hence

$$\begin{aligned} p^{2^{k-1}}(1-p)^{2^{k-1}-1}P_{\text{ue}}(C,p) &= P_{\text{ue}}(C^*,p) \\ &\geq P_{\text{ue}}(C_1,p) = p^{2^{k-1}}(1-p)^{2^{k-1}-1}P_{\text{ue}}(C_1^-,p) \end{aligned}$$

and

$$P_{\text{ue}}(C,p) \leq P_{\text{ue}}(C_1^-,p)$$

and so $P_{\text{ue}}(C^*,p) = P_{\text{ue}}(C_1,p)$, that is, C^* is optimal for p. This proves (i), and (ii) is similar.

By Lemma 3.6.4, for $k \leq 4$ it is sufficient to determine the optimal $[n,k]$ codes for $p \in \left(0,\frac{1}{2}\right)$ where $n < 2^k - 1 + k$. This can be done by a computer search. The search gives the following result.

Theorem 3.6.2 *If $1 \leq k \leq 4$ and $n \geq k$, then there exists an $[n,k]$ code which is optimal for all $p \in \left[0,\frac{1}{2}\right]$.*

Column count functions for optimal codes are given in Tables 3.1-3.3.

n_0	$\bar{x}$	$A_0(z)$
	101	
	011	
0	000	4
1	100	$2+2z$
2	110	$1+2z+z^2$

Table 3.1 $m_0(\bar{x})$ and $A_0(z)$ for $k=2$.

n_0	$\bar{x}$	$A_0(z)$
	1010101 0110011 0001111	
0	0000000	8
1	1000000	$4+4z$
2	1100000	$2+4z+2z^2$
3	1101000	$1+3z+3z^2+z^3$
4	1101001	$1+6z^2+z^4$
5	1111100	$1+2z^2+4z^3+z^4$
6	1111110	$1+4z^3+3z^4$

Table 3.2 $m_0(\bar{x})$ and $A_0(z)$ for $k=3$.

n_0	$\bar{x}$	$A_0(z)$
	101010101010101 011001100110011 000111100001111 000000011111111	
0	000000000000000	16
1	100000000000000	$8+8z$
2	110000000000000	$4+8z+4z^2$
3	110100000000000	$2+6z+6z^2+2z^3$
4	110100010000000	$1+4z+6z^2+4z^3+z^4$
5	110100010000001	$1+10z^2+5z^4$
6	110100110010000	$1+3z^2+8z^3+3z^4+z^6$
7	110100110010100	$1+7z^3+7z^4+z^7$
8	110100110010110	$1+14z^4+z^8$
9	111110011000011	$1+6z^4+8z^5+z^8$
10	111111011000011	$1+2z^4+8z^5+4z^6+z^8$
11	111111011100110	$1+6z^5+6z^6+2z^7+z^8$
12	111111011100111	$1+12z^6+3z^8$
13	111111111111100	$1+4z^6+8z^7+3z^8$
14	111111111111110	$1+8z^7+7z^8$

Table 3.3 $m_0(\bar{x})$ and $A_0(z)$ for $k=4$.

We write $n = r(2^k - 1) + n_0$ where $0 \le n_0 \le 2^k - 1$. Then the column count function for an optimal code is given by

$$m_C(\bar{x}) = r + m_0(\bar{x})$$

and

$$A_C(z) - 1 = z^{r2^{k-1}}\{A_0(z) - 1\}$$

where m_0 and A_0 are given in the tables.

Copies of the Simplex Code

Let k and s by any positive integers. Let C be the $[s(2^k - 1), k]$ code generated by the matrix G which contains as columns s copies of each non-zero vector in $GF(2)^k$. This code has minimum distance $d = s2^{k-1}$. From Theorems 3.5.1 and 3.5.8 we get

$$P_{\text{ue}}(C,p) = (2^k - 1)p^d(1-p)^{n-d} = P[s(2^k - 1), k, p].$$

Hence the code C is optimal. In particular, for $s = 1$ we get the simplex code whose dual is the Hamming code. Hence the Hamming codes are also optimal.

3.7 NEW CODES FROM OLD

As illustrated in the previous chapter, there are a number of ways to construct new codes from one or two old ones. For many of these construction, the new code may not be good even if the old is/are. We shall give a number of examples which illustrates this. The *-operation is atypical in this respect. It gives a systematic way to make long good codes and we discuss this construction first.

The *-operation

Theorem 3.7.1 *(i) If C is good, then C^* is good.*
(ii) If C is proper, then C^* is proper.

Proof: Since $p^{2^{k-1}}(1-p)^{2^{k-1}-1}$ is increasing on $[0, \frac{1}{2}]$ and

$$P_{\text{ue}}(C^*,p) = p^{2^{k-1}}(1-p)^{2^{k-1}-1}P_{\text{ue}}(C,p),$$

(ii) follows. Further,

$$p^{2^{k-1}}(1-p)^{2^{k-1}-1} \leq \left(\frac{1}{2}\right)^{2^k-1},$$

and so *(i)* follows.

Theorem 3.7.1 shows that we can make longer good codes from shorter ones. However, there is a stronger result which shows that starting from any code and using the *-operation repeatedly, we will eventually get a good code, and even a proper code.

Theorem 3.7.2 *If C is an $[n,k]$ code and $r \geq n-k$, then C^{r*} is good.*

Proof: By Theorem 3.5.10, $P_{\text{ue}}(C,p) \leq (1-p)^{n-k}\left\{1-(1-p)^k\right\}$. Hence

$$\begin{aligned}
P_{\text{ue}}(C^{(n-k)*},p) &= \left\{p^{2^{k-1}}(1-p)^{2^{k-1}-1}\right\}^{n-k} P_{\text{ue}}(C,p) \\
&\leq \left\{p^{2^{k-1}}(1-p)^{2^{k-1}-1}\right\}^{n-k} (1-p)^{n-k}\left\{1-(1-p)^k\right\} \\
&= \{p(1-p)\}^{(n-k)2^{k-1}}\left\{1-(1-p)^k\right\} \\
&\leq \left(\frac{1}{4}\right)^{(n-k)2^{k-1}}\left\{1-\left(\frac{1}{2}\right)^k\right\} \\
&= \frac{2^k-1}{2^{n+(n-k)(2^k-1)}}.
\end{aligned}$$

since $p(1-p)$ and $\left\{1-(1-p)^k\right\}$ are monotonously increasing on $\left[0,\frac{1}{2}\right]$. This proves the theorem for $r=n-k$, and the general case then follows from Theorem 3.7.1(i) by induction.

Theorem 3.7.3 *If C is an $[n,k,d]$ code and $r \geq \max\{0, n-2d\}$, then C^{r*} is proper.*

Proof: By Theorem 3.1.4, $P_{\text{ue}}(C,p)$ is monotonously increasing on $\left[0,\frac{d}{n}\right]$. Hence, if $\frac{d}{n} \geq \frac{1}{2}$, i.e. $2d \geq n$, then the theorem follows. If $2d < n$, then $C^{(n-2d)*}$ is an $[n+(n-2d)(2^k-1), k, d+(n-2d)2^{k-1}]$ code and

$$\frac{d+(n-2d)2^{k-1}}{n+(n-2d)(2^k-1)} = \frac{1}{2}.$$

Hence, $C^{(n-2d)*}$ is proper. The general result follows from Theorem 3.7.1(ii) by induction.

Corollary 3.7.1 *If C is an $[n,k]$ code and $r \geq n-2$, then C^{r*} is proper.*

Adding/Removing a Parity Bit

Consider an $[n,k]$ code C. Adding a parity bit, that is, extending each code word $\bar{a}$ to $(\bar{a}|\sum_{i=1}^{n} a_i)$ gives an $[n+1,k]$ code C^{ex} where all the code words have even weight. The weight distribution function of C^{ex} is given by

$$A_{C^{\text{ex}}}(z) = \sum_{i=0}^{\lfloor \frac{n+1}{2} \rfloor} (A_{2i-1} + A_{2i}(C))z^{2i} = \frac{1}{2}\{A_C(z) + A_C(-z)\}.$$

The code C may be good without C^{ex} being good and vice versa. Examples of the various combinations of possibilities are given in Figure 3.6 which gives a generator matrix and the weight distribution function of the corresponding code as well as a plot of the corresponding $P(C,p)/P(C,0.5)$. Puncturing the last position of C^{ex} we get C. From the examples in Figure 3.6 the see that a code may be good without the punctured code being good and vice versa.

Even-Weight Subcodes

Consider an $[n,k]$ code C containing some code word of odd weight. The even-weight $[n,k-1]$ subcode C^{e} is the set of code words in C of even weight. The weight distribution function of C^{e} is given by

$$A_{C^{\text{e}}}(z) = \sum_{i=0}^{\lfloor \frac{n}{2} \rfloor} A_{2i}(C)z^{2i}.$$

We illustrate the various possibilities in Figure 3.7.

	C^{ex} proper	C^{ex} not good
C proper	$\begin{pmatrix} 10 \\ 01 \end{pmatrix}$	$\begin{pmatrix} 1000001 \\ 0100001 \\ 0011111 \end{pmatrix}$
$A_C(z)$:	$1+2z+z^2$	$1+3z^2+3z^5+z^7$
$A_{C^{\text{ex}}}(z)$:	$1+3z^2$	$1+3z^2+3z^6+z^8$
C not good	$\begin{pmatrix} 100 \\ 010 \end{pmatrix}$	$\begin{pmatrix} 1000 \\ 0100 \end{pmatrix}$
$A_C(z)$:	$1+2z+z^2$	$1+2z+z^2$
$A_{C^{\text{ex}}}(z)$:	$1+3z^2$	$1+3z^2$

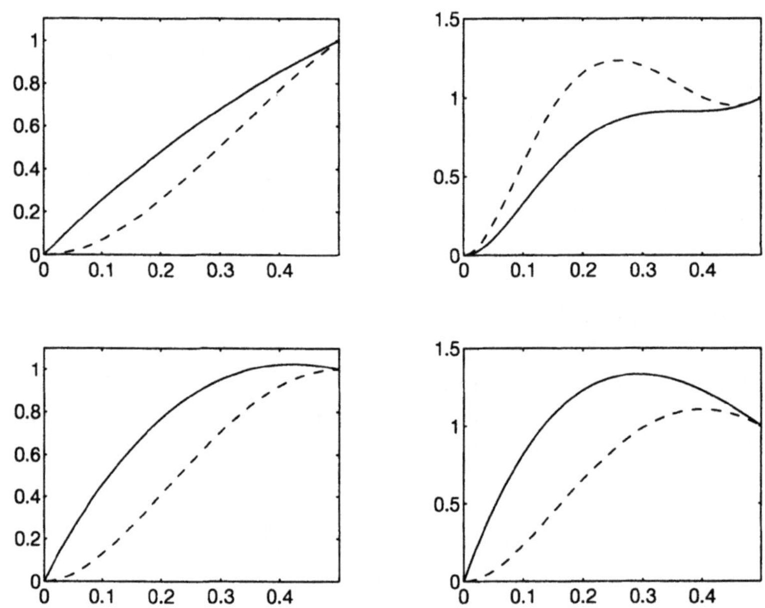

Figure 3.6 Plot of extended code examples; $P_{\text{ue}}(C,p)/P_{\text{ue}}(C,0.5)$ (solid line) and $P_{\text{ue}}(C^{\text{ex}},p)/P_{\text{ue}}(C^{\text{ex}},0.5)$ (dashed line).

	C^{e} proper	C^{e} not good
C proper	$\begin{pmatrix} 1000 \\ 0100 \\ 0011 \end{pmatrix}$	$\begin{pmatrix} 11000 \\ 01100 \\ 00111 \end{pmatrix}$
$A_C(z)$:	$1 + 2z + 2z^2 + 2z^3 + z^4$	$1 + 3z^2 + 3z^3 + z^5$
$A_{C^{\mathrm{e}}}(z)$:	$1 + 2z^2 + z^4$	$1 + 3z^2$
C not good	$\begin{pmatrix} 1000 \\ 0100 \\ 0010 \end{pmatrix}$	$\begin{pmatrix} 10000 \\ 01000 \\ 00100 \end{pmatrix}$
$A_C(z)$:	$1 + 3z + 3z^2 + z^3$	$1 + 3z + 3z^2 + z^3$
$A_{C^{\mathrm{e}}}(z)$:	$1 + 3z^2$	$1 + 3z^2$

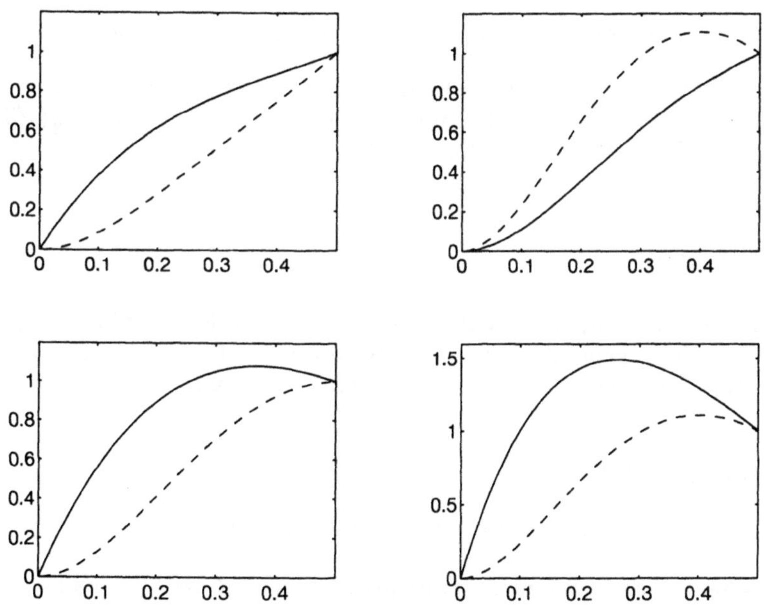

Figure 3.7 Examples of even-weight subcodes, and plots of $P_{\mathrm{ue}}(C,p)/P_{\mathrm{ue}}(C,0.5)$ (solid line) and $P_{\mathrm{ue}}(C^{\mathrm{e}},p)/P_{\mathrm{ue}}(C^{\mathrm{e}},0.5)$ (dashed line).

Shortened Codes

Consider a systematic $[n,k]$ code C and let C^s be the shortened $[n-1,k-1]$ code:

$$C^s = \left\{ \bar{x} \in GF(2)^{n-1} \,\middle|\, (0|\bar{x}) \in C \right\}.$$

Again the shortened code may be good or not good, whether the original code is good or not, as shown by the examples in Figure 3.8. We note that

$$P\left(C^s, \frac{1}{2}\right) = \frac{2^{k-1}-1}{2^{n-1}} < \frac{2^k-1}{2^n} = P\left(C, \frac{1}{2}\right).$$

It is natural to ask if $P(C^s,p) \le P(C,p)$ for all $p \in \left[0, \frac{1}{2}\right]$. In terms of the weight distribution this can be written as

$$\Delta_C(z) \stackrel{\text{def}}{=} A_C(z) - 1 - (1+z)\left\{A_{C^s}(z) - 1\right\} \ge 0$$

for all $z \in [0,1]$. The examples in Figure 3.8 show that this may or may not be true. Clearly $z \ge 0$ for $z \in [0,1]$. But, $-z^2+z^3+z^4 < 0$ for $z \in [0, \frac{1}{2}(\sqrt{5}-1)]$, and $-2z^2 - z^3 + 2z^4 + 2z^5 < 0$ for $z \in [0, 0.87555]$.

Product Codes

Let C_1 be an $[n_1,k_1,d_1]$ code and C_2 an $[n_2,k_2,d_2]$ code. Let C be the product code. There is in general no simple expression for $A_C(z)$ in terms of $A_{C_1}(z)$ and $A_{C_2}(z)$. Here we consider a more general situation.

Theorem 3.7.4 *Let*

$$\mathcal{C}_1 = \{C_{1i} \mid i \ge 1\} \text{ and } \mathcal{C}_2 = \{C_{2i} \mid i \ge 1\}$$

be two classes of codes where C_{1i} is an $[n_{1i}, k_{1i}, d_1]$ code and C_{2i} an $[n_{2i}, k_{2i}, d_2]$ code, where $n_{ji} > k_{ji}$ for $j = 1,2$, $i \ge 1$. Let C_i be the product of C_{1i} and C_{2i}. Then C_i is not good when $n_{1i}n_{2i}$ is sufficiently large.

Proof: First we note that

$$n_{1i}n_{2i} - k_{1i}k_{2i} \ge n_{1i}n_{2i} - (n_{1i}-1)(n_{2i}-1) = n_{1i} + n_{2i} - 1.$$

	C^s proper	C^s not good
C proper	$\begin{pmatrix} 100 \\ 010 \\ 001 \end{pmatrix}$	$\begin{pmatrix} 1001 \\ 0100 \\ 0010 \end{pmatrix}$
$A_C(z)$:	$1 + 3z + 3z^2 + z^3$	$1 + 2z + 2z^2 + 2z^3 + z^4$
$A_{C^s}(z)$:	$1 + 2z + z^2$	$1 + 2z + z^2$
$\Delta_C(z)$:	z	$-z^2 + z^3 + z^4$
C not good	$\begin{pmatrix} 1000000 \\ 0101011 \\ 0010111 \end{pmatrix}$	$\begin{pmatrix} 100111 \\ 010000 \\ 001000 \end{pmatrix}$
$A_C(z)$:	$1 + z + 3z^4 + 3z^5$	$1 + 2z + z^2 + z^4 + 2z^5 + z^6$
$A_{C^s}(z)$:	$1 + 3z^4$	$1 + 2z + z^2$
$\Delta_C(z)$:	z	$-2z^2 - z^3 + 2z^4 + 2z^5$

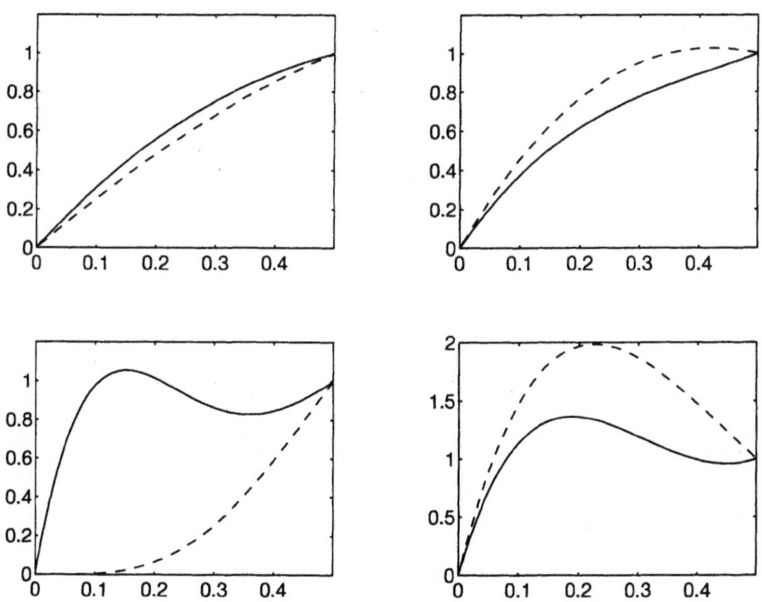

Figure 3.8 Examples, with plots of $P_{\text{ue}}(C,p)/P_{\text{ue}}(C,0.5)$ (solid line) and $P_{\text{ue}}(C^s,p)/P_{\text{ue}}(C^s,0.5)$ (dashed line), of some shortened codes.

Rough estimates give

$$\begin{aligned}\frac{P(C_i,\frac{d_1d_2}{n_{1i}n_{2i}})}{P(C_i,\frac{1}{2})} &\geq 2^{n_{1i}n_{2i}-k_{1i}k_{2i}}\left(\frac{d_1d_2}{n_{1i}n_{2i}}\right)^{d_1d_2}\left(1-\frac{d_1d_2}{n_{1i}n_{2i}}\right)^{n_{1i}n_{2i}-d_1d_2}\\ &> 2^{n_{1i}+n_{2i}-1}\left(\frac{d_1d_2}{n_{1i}n_{2i}}\right)^{d_1d_2}\left(1-\frac{d_1d_2}{n_{1i}n_{2i}}\right)^{n_{1i}n_{2i}}\\ &\geq 2^{n_{1i}+n_{2i}-1}(d_1d_2)^{d_1d_2}\frac{1}{(n_{1i}n_{2i})^{d_1d_2}}\left(\frac{1}{4}\right)^{d_1d_2}\end{aligned}$$

if $n_{1i}n_{2i} \geq 2d_1d_2$. Since $2^{n_{1i}+n_{2i}-1}$ is exponential in $\max\{n_{1i}, n_{2i}\}$ whereas $(n_{1i}n_{2i})^{d_1d_2}$ is polynomial, the theorem follows.

3.8 COMBINED CORRECTION AND DETECTION

Using a Single Code for Correction and Detection

In 2.4 we showed that the probability of undetected error after using the (n, M, d) code C to correct t or less errors, where $2t + 1 \leq d$, is given by

$$P_{\text{ue}}^{(t)}(C,p) = \sum_{j=d-t}^{n} A_{t,j}p^j(1-p)^{n-j} \tag{3.18}$$

where

$$A_{t,j} = \sum_{i=j-t}^{j+t} A_i N_t(i,j)$$

and $N_t(i,j)$ is given by (2.10).

Any question considered for $P_{\text{ue}}(C,p)$ may be considered for $P_{\text{ue}}^{(t)}(C,p)$ for $t > 0$, e.g. bounds, means, etc. In particular, we call C *t-proper* if $P_{\text{ue}}^{(t)}(C,p)$ is monotonous and *t-good* if

$$P_{\text{ue}}^{(t)}(C,p) \leq P_{\text{ue}}^{(t)}\left(C,\frac{1}{2}\right) \tag{3.19}$$

for all $p \in \left[0, \frac{1}{2}\right]$.

In Figure 3.9 we give examples of generator matrices of two $[12, 4]$ codes which show that a code C may be proper without being 1-proper, or good without being 1-good.

C is proper is not 1-proper	$\begin{pmatrix} 100011110000 \\ 010001101111 \\ 001000010001 \\ 000100010011 \end{pmatrix}$
$A_C^{(0)}(z)$: $A_C^{(1)}(z)$:	$2z^3 + z^4 + z^5 + 2z^6 + 4z^7 + 4z^8 + z^9$ $6z^2 + 6z^3 + 24z^4 + 21z^5 + 37z^6$ $+48z^7 + 33z^8 + 17z^9 + 3z^{10}$
C is good is not 1-good	$\begin{pmatrix} 100011100101 \\ 010000011001 \\ 001000011111 \\ 000100011011 \end{pmatrix}$
$A_C^{(0)}(z)$: $A_C^{(1)}(z)$:	$2z^3 + 2z^4 + z^5 + 2z^6 + 2z^7 + 3z^8 + 3z^9$ $6z^2 + 10z^3 + 25z^4 + 29z^5 + 23z^6$ $+38z^7 + 40z^8 + 15z^9 + 9z^{10}$

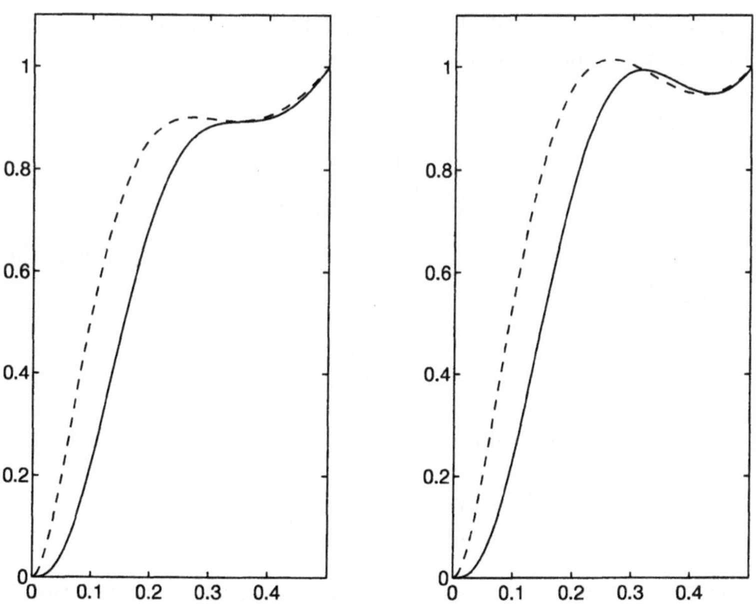

Figure 3.9 Examples of two $[12, 4]$ codes. Generator matrices are given above and and plots showing $P_{\text{ue}}^{(0)}(C, p) / P_{\text{ue}}^{(0)}(C, 0.5)$ (solid line) and $P_{\text{ue}}^{(1)}(C, p) / P_{\text{ue}}^{(1)}(C, 0.5)$ (dashed line) below.

A Simple Upper Bound

Let C be an $[n,k,d]$ code where $d \geq 2t+1$. When $\bar{0}$ is sent and $\bar{y}$ is received, then we have an undetected error, after correcting of t or less errors, if and only of $\bar{y} \in \bigcup_{\bar{x}\in C\setminus\{\bar{0}\}} S_t(\bar{x})$. By definition, the spheres $S_t(\bar{x})$ are disjoint and the union is a subset of $GF(2)^n$. Hence $A_{t,j} \leq \binom{n}{j}$ and we get the following upper bound.

Theorem 3.8.1 *Let C be an $[n,k,d]$ code with $d \geq 2t+1$. Then*

$$P_{\text{ue}}^{(t)}(C,p) \leq \sum_{j=d-t}^{n} \binom{n}{j} p^j (1-p)^{n-j}.$$

The Ratio $P_{\text{ue}}^{(t)}(C,p)/P_{\text{ue}}(C,p)$

We now consider how the probability of undetected error increases when some of the power of the code is used for error correction.

For $p = \frac{1}{2}$ we have

$$\frac{P_{\text{ue}}^{(t)}(C,\frac{1}{2})}{P_{\text{ue}}(C,\frac{1}{2})} = \sum_{j=0}^{t} \binom{n}{j}.$$

For small values of p, the dominating term in $P_{\text{ue}}(C,p)$ is $A_d p^d$ and the dominating term in $P_{\text{ue}}^{(t)}(C,p)$ is

$$A_{t,d-t} p^{d-t} = \binom{d}{t} A_d p^{d-t}.$$

That is, the probability of undetected error increases by a factor

$$\frac{P_{\text{ue}}^{(t)}(C,p)}{P_{\text{ue}}(C,p)} = \binom{d}{t} p^{-t} Q_t(C,p),$$

where $Q_t(C,p) \to 1$ when $p \to 0+$. To get more precise information, we can consider the Taylor expansion

$$Q_t(C,p) = \sum_{i=0}^{\infty} \alpha_i p^i.$$

The coefficients α_i will in general be complicated expressions involving the A_j-s. In particular,

$$\begin{aligned}
\alpha_0 &= 1,\\
\alpha_1 &= \frac{t}{d-t+1}\left(-d+t+\frac{A_{d+1}}{A_d}\right),\\
\alpha_2 &= \frac{t(t-1)(d-t)}{2(d-t+2)}+\frac{t(n-d)}{d-t+1}+\frac{A_{d+1}t}{A_d(d-t+1)}\left(1-\frac{t(d-t+1)+1}{d-t+2}\right)\\
&\quad -\frac{A_{d+1}^2 t}{A_d^2(d-t+1)}+\frac{A_{d+2}(2d+3-t)}{A_d(d-t+1)(d-t+2)}.
\end{aligned}$$

Average

One can find average results for $P_{\mathrm{ue}}^{(t)}(p)$ similar to those found for $P_{\mathrm{ue}}(p)$ in section 3.2. The proof is also similar and we omit the details. The result can be expressed in various ways, and we list several.

Theorem 3.8.2 *Let $\mathcal{C}$ be a set of codes of length n and minimum distance at least $d = 2t+1$. Then*

$$\begin{aligned}
P_{\mathrm{ue}}^{(t)}(\mathcal{C},p) &= \sum_{w_H(\bar{x})\ge 2t+1} \alpha(\bar{x}) \sum_{\bar{y}\in S_t(\bar{x})} p^{w_H(\bar{y})}(1-p)^{n-w_H(\bar{y})}\\
&= \sum_{i=2t+1}^{n} A_i(\mathcal{C}) \sum_{j=i-t}^{i+t} N_t(i,j)p^j(1-p)^{n-j}\\
&= \sum_{w_H(\bar{y})\ge t+1} p^{w_H(\bar{y})}(1-p)^{n-w_H(\bar{y})} \sum_{\bar{x}\in S_t(\bar{y})} \alpha(\bar{x}).
\end{aligned}$$

We consider one example.

Theorem 3.8.3 *For the set $\mathcal{C} = \mathcal{SYS}(n,k,2t+1)$ we have*

$$P_{\mathrm{ue}}^{(t)}(\mathcal{C},p) \le \frac{\sum_{i=0}^{t}\binom{n}{i}}{2^{n-k}-\sum_{i=0}^{d-2}\binom{n-1}{i}}\left\{1-\sum_{i=0}^{t}\binom{n}{i}p^i(1-p)^{n-i}\right\}.$$

Proof: Let

$$\beta = \frac{1}{2^{n-k}-\sum_{i=0}^{d-2}\binom{n-1}{i}}.$$

By (3.4) we have $\alpha(\bar{x}) \leq \beta$ for all $\bar{x}$ of weight at least $2t+1$. Hence

$$\begin{aligned} P_{\text{ue}}^{(t)}(C,p) &\leq \beta \sum_{w_H(\bar{y})\geq t+1} p^{w_H(\bar{y})}(1-p)^{n-w_H(\bar{y})} \# S_t(\bar{y}) \\ &= \beta\left\{1-\sum_{i=0}^{t}\binom{n}{i}p^i(1-p)^{n-i}\right\}\sum_{i=0}^{t}\binom{n}{i}. \end{aligned}$$

Concatenated Codes for Error Correction and Detection

Two types of concatenated codes for combined error correction and detection have been studied.

$m = 1$

Let the outer code D be an $[k,l]$ code and the inner code C an $[n,k]$ systematic code. Let $u(\bar{x}) \in GF(2)^{n-k}$ be the check bits corresponding to the information vector $\bar{x} \in GF(2)^k$, and let

$$E = \{(\bar{x}|u(\bar{x})) \mid \bar{x} \in D\}.$$

Then E is an $[n,l]$ code. Suppose without loss of generality that the all-zero code word is sent. At the receiving end we first use complete maximum likelihood decoding for the code C, that is we decode into the closest code word of $(\bar{x}|u(\bar{x})) \in C$, and if there is more then one closest code word, we choose one of these at random. If the corresponding $\bar{x}$ is a non-zero code word in D, then we have an undetectable error after decoding. Let $P_{\text{ue}}(D,C,p)$ be the probability that this happens. The question we must ask, how does use of the code C affect the probability of undetected error, that is, what is

$$R(D,C,p) \stackrel{\text{def}}{=} \frac{P_{\text{ue}}(D,C,p)}{P_{\text{ue}}(D,p)}.$$

By a careful choice of codes it is possible to get $R(D,C,p) \to 0$ when $p \to 0$. In particular, this is the case if $d_{\min}(C) > 2d_{\min}(D)$.

Remark. An alternative way of using the combined code is to correct up to t errors using code C and the use the remaining power to detect errors; this is a special case of the next method considered. In this case an error is undetected if the received vector is within distance t of a code word of the form $(\bar{x}|u(\bar{x}))$, where $\bar{x} \in D \setminus \{\bar{0}\}$, that is, within distance t of a non-zero code word in E. Hence the probability of undetected error is $P_{\text{ue}}^{(t)}(E,p)$.

$m \geq 1$

Let D be an $[mk, K]$ code and C an $[n, k]$ code which is able to correct t errors. Let E denote the concatenated code. Suppose that the all zero vector in $GF(q)^{mn}$ is transmitted. Let $(\bar{y}_1, \bar{y}_2, \cdots, \bar{y}_m)$ be received. The probability that $\bar{y}_l$ is within distance t of a fixed code word in C of weight i is given by

$$p'(i) = \sum_{j=0}^{n} N_t(i,j)p^j(1-p)^{n-j}.$$

If some $\bar{y}_l$ is at distance more then t from all code words in C, then a detectable error has occurred. Otherwise, each $\bar{y}_l$ is decoded to the closest code word $\tilde{y}_l \in GF(q)^n$.

Let the corresponding information vector be $\tilde{z}_l \in GF(q)^k$. If $(\tilde{z}_1|\tilde{z}_2|\cdots|\tilde{z}_m)$ is a non-zero code word in D, then we have an undetectable error. The probability that this happens is then given by

$$\sum_{(i_1,i_2,\cdots,i_m)\neq(0,0,\cdots,0)} A_{i_1,i_2,\cdots,i_m}(C)p'(i_1)p'(i_2)\cdots p'(i_m).$$

3.9 COMPLEXITY OF COMPUTING P_{UE}

The following theorem shows that it is a hard problem to compute $P_{\text{ue}}(C,p)$ in general. For a generator matrix G, let C_G denote the code generated by G.

Theorem 3.9.1 *The problem of computing $P_{\text{ue}}(C_G, p)$, as a function of a rational p and a generator matrix G, is an $\mathcal{NP}$ hard problem.*

Proof: It is known [37] that the following problem is $\mathcal{NP}$ complete:

> Given a $k \times n$ (binary) generator matrix G and an integer w, decide if the code C_G contains a code word of weight w.

In particular, this implies that the problem of finding the weight distribution of C_G is $\mathcal{NP}$ hard. We show that the problem of finding the weight distribution of C_G has the same complexity as the problem of evaluating $P_{\text{ue}}(C_G, p)$ in the sense that each has a polynomial time reduction to the other.

First, computing $P_{\text{ue}}(C_G, p)$, given a rational p and the weight distribution of C_G, is a simple evaluation of a polynomial using Theorem 2.3.1, and this can be done in polynomial time.

Next, if we know $P_{\text{ue}}(C_G, p_i)$ for n different values $p_1, p_2, \ldots, p_n$ (all different from 1), then the weight distribution can be determined in polynomial time from the set of n linear equation:

$$\begin{aligned} z_1 A_1 + z_1^2 A_2 + \cdots + z_1^n A_n &= (1 - p_1)^n P_{\text{ue}}(C_G, p_1), \\ z_2 A_1 + z_2^2 A_2 + \cdots + z_2^n A_n &= (1 - p_2)^n P_{\text{ue}}(C_G, p_2), \\ \ldots \\ z_n A_1 + z_n^2 A_2 + \cdots + z_n^n A_n &= (1 - p_n)^n P_{\text{ue}}(C_G, p_n), \end{aligned}$$

where $z_i = p_i/(1 - p_i)$. Since the coefficient matrix of this system of equations is a Vandermonde matrix, if has full rank and the set of equations determine $A_1, A_2, \ldots A_n$ uniquely.

3.10 PARTICULAR CODES

In this section we list the weight distributions of some known classes of codes. In most cases we refer only to binary codes, but some important classes of non-binary code are also considered.

3.10.1 Perfect Codes

Repetition Codes and Their Duals

Over $GF(q)$ the (generalized) repetition codes are the $[n, 1; q]$ codes whose non-zero code words have Hamming weight n. The weight distribution function is

$$1 + (q - 1)z^n.$$

Both the code and its dual are proper. In the binary case, the dual code is known as the single parity check code.

Hamming and Simplex Codes

For given m and q, the simplex code is the cyclic code over $GF(q)$ generated by a primitive polynomial of degree m. It has dimension m and length $n = \frac{q^m - 1}{q - 1}$.

All non-zero code words have weight q^{m-1}, that is, its weight distribution function is

$$1 + (q^m - 1)\, z^{q^{m-1}}.$$

The dual code is the $[n, n-m]$ Hamming code. The weight distribution can be found from the weight distribution of the simplex code, using MacWilliams' identity:

$$A_i = \binom{\frac{q^m-1}{q-1}}{i} \frac{(q-1)^i}{q^m} + \frac{q^m-1}{q^m} \sum_{l=0}^{i} \binom{q^{m-1}}{l} \binom{\frac{q^{m-1}-1}{q-1}}{i-l} (-1)^l (q-1)^{i-l}.$$

Both codes are proper. The extended Hamming code and its dual code are also both proper.

Estimates of the minimum distances of shortened binary simplex codes (the partial periods of binary m-sequences) were given by Kumar and Wei [124]. Whether such shortened codes are good or not will depend on the primitive polynomial used for generating the code and the shortening done. Shortening of Hamming codes will be considered in more detail below (CRC codes).

Golay Codes

The binary Golay code is the $[23, 12, 7]$ CRC code generated by the polynomial

$$z^{11} + z^{10} + z^6 + z^5 + z^4 + z^2 + 1.$$

Its weight distribution is

i:	0	7	8	11	12	15	16	23
A_i:	1	253	506	1288	1288	506	253	1

Both the code and its dual are proper.

The ternary Golay code is the $[11, 6, 5; 3]$ CRC code generated by the polynomial

$$z^5 + z^4 - z^3 + z^2 - 1.$$

Its weight distribution is

i:	0	5	6	8	9	11
A_i:	1	132	132	330	110	24

Both the code and its dual are proper.

References: [13, p. 69, 482], Leung et al. [133], Mallow et al. [144].

3.10.2 Reed-Muller Codes

The r'th order Reed-Muller code of length 2^m is a binary $\left[2^m, \sum_{i=0}^{r} \binom{m}{i}\right]$ code. The first order Reed-Muller code is the dual of the extended Hamming code and has weight distribution

i	A_i
0	1
2^{m-1}	$2^m - 2$
2^m	1

The code is proper. The weight distribution of the second order Reed-Muller code was determined by Sloane and Berlekamp, see [13, p. 443]:

i	A_i
0	1
$2^{m-1} \pm 2^{m-1-h}$	$2^{h(h+1)} \frac{\prod_{i=m-2h+1}^{m}(2^i-1)}{\prod_{i=1}^{h}(2^{2i}-1)}$ for $1 \le h \le \lfloor \frac{m}{2} \rfloor$
2^m	1

The remaining code words have weight 2^{m-1}, that is,

$$A_{2^{m-1}} = 2^{(m^2+m+2)/2} - 2 - 2 \sum_{h=1}^{\lfloor \frac{m}{2} \rfloor} 2^{h(h+1)} \frac{\prod_{i=m-2h+1}^{m} (2^i - 1)}{\prod_{i=1}^{h} (2^{2i} - 1)}.$$

The code is proper for $m \le 5$, but it is not good for $m \ge 6$. The class of second order Reed-Muller codes is asymptotically bad. In fact, any infinite subset of the set $\{R(r, m) \mid r \ge 2, m \ge r + 3\}$ is asymptotically bad, see Kløve [109].

Kasami [90] determined the weight distribution of several subcodes of the second order Reed-Muller codes.

3.10.3 BCH Codes

The primitive BCH codes are defined as follows. Let α be a primitive element of $GF(q^m)$. Let $M_j(x)$ be the polynomial of lowest degree over $GF(q)$ having α^j as a root (the *minimal polynomial* of α^j). The t-error correcting BCH code (for short: t-BCH code) is the CRC code generated by the polynomial

$$\operatorname{lcm}\{M_1(x), M_2(x), \ldots, M_{2t}(x)\}.$$

The code has length $n = (q^m - 1)/(q - 1)$, minimum distance at least $2t + 1$ and dimension at least $n - tm$. The 1-BCH code is the Hamming code.

The weight distribution of the dual code of the binary 2-BCH is given by

i	A_i
0	1
$2^{m-1} - 2^{(m-1)/2}$	$(2^m - 1)\left(2^{m-2} + 2^{(m-3)/2}\right)$
2^{m-1}	$(2^m - 1)\left(2^{m-1} + 1\right)$
$2^{m-1} + 2^{(m-1)/2}$	$(2^m - 1)\left(2^{m-2} - 2^{(m-3)/2}\right)$

for m odd and

i	A_i
0	1
$2^{m-1} - 2^{m/2}$	$2^{(m-4)/2}\left(2^m - 1\right)\left(2^{(m-2)/2} + 1\right)/3$
$2^{m-1} - 2^{m/2-1}$	$2^{m/2}\left(2^m - 1\right)\left(2^{m/2} + 1\right)/3$
2^{m-1}	$(2^m - 1)\left(2^{m-2} + 1\right)$
$2^{m-1} + 2^{m/2-1}$	$\frac{1}{3}2^{m/2}\left(2^m - 1\right)\left(2^{m/2} - 1\right)$
$2^{m-1} + 2^{m/2}$	$\frac{1}{3}2^{(m-4)/2}\left(2^m - 1\right)\left(2^{(m-2)/2} - 1\right)$

for m even, see [13, p. 451f] or [11, p. 177f]. Leung et al. [133] proved that the binary 2-BCH and the extended binary 2-BCH are both proper.

The binary 3-BCH code is a $[2^m - 1, 2^m - 1 - 3m]$ code whose dual code has the weight distribution

i	A_i
0	1
$2^{m-1} - 2^{(m+1)/2}$	$2^{(m-5)/2}\left(2^m - 1\right)\left(2^{m-1} - 1\right)\left(2^{(m-3)/2} + 1\right)/3$
$2^{m-1} - 2^{(m-1)/2}$	$2^{(m-3)/2}\left(2^m - 1\right)\left(5 \cdot 2^{m-1} - 1\right)\left(2^{(m-1)/2} + 1\right)/3$
2^{m-1}	$(2^m - 1)\left(9 \cdot 2^{2m-4} + 3 \cdot 2^{m-3} + 1\right)$
$2^{m-1} + 2^{(m-1)/2}$	$2^{(m-3)/2}\left(2^m - 1\right)\left(5 \cdot 2^{m-1} - 1\right)\left(2^{(m-1)/2} - 1\right)/3$
$2^{m-1} + 2^{(m+1)/2}$	$2^{(m-5)/2}\left(2^m - 1\right)\left(2^{m-1} - 1\right)\left(2^{(m-3)/2} - 1\right)/3$

for $m \geq 5$, m odd and

i	A_i
0	1
$2^{m-1} - 2^{(m+2)/2}$	$(2^m - 1)(2^m - 4)\left(2^{m-1} + 2^{(m+2)/2}\right)/960$
$2^{m-1} - 2^{m/2}$	$7(2^m - 1)2^m\left(2^{m-1} + 2^{m/2}\right)/48$
$2^{m-1} - 2^{(m-2)/2}$	$(2^m - 1)\left(2^{m-1} + 2^{(m-2)/2}\right)(6 \cdot 2^m + 16)/15$
2^{m-1}	$(2^m - 1)\left(29 \cdot 2^{2m} - 2^{m+2} + 64\right)/64$
$2^{m-1} + 2^{(m-2)/2}$	$(2^m - 1)\left(2^{m-1} - 2^{(m-2)/2}\right)(6 \cdot 2^m + 16)/15$
$2^{m-1} + 2^{m/2}$	$7(2^m - 1)2^m\left(2^{m-1} - 2^{m/2}\right)/48$
$2^{m-1} + 2^{(m+2)/2}$	$(2^m - 1)(2^m - 4)\left(2^{m-1} - 2^{(m+2)/2}\right)/960$

for $m \geq 6$, m even, see [13, p. 669] and [11, p. 178]. Ong and Leung [155] proved that for m odd the 3-BCH and the corresponding extended code are both proper. For m even, however, Perry [158] showed that neither the 3-BCH nor the extended 3-BCH are good for $m \geq 6$. However, these classes of codes are asymptotically good.

3.10.4 MDS and Almost MDS codes

The Singelton bound says that $d \leq n - k + 1$ for an $[n, k, d; q]$ code. Maximum distance separable (MDS) codes are $[n, k, d; q]$ codes where $d = n - k + 1$. For these codes

$$A_i = \binom{n}{i} \sum_{j=0}^{i-n+k-1} (-1)^j \binom{i}{j} \left(q^{i-j-n+k} - 1\right)$$

for $n - k + 1 \leq i \leq n$, see [16] p.72. Kasami and Lin [95] proved that $P_{\mathrm{ue}}^{(t)}(C, p)$ is monotonous on $[0, \frac{q-1}{q}]$ for all $[n, k, d; q]$ MDS codes C and all $t < \frac{d}{2}$, that is, the codes are t-proper.

An $[n, k; 2^m]$ code can also be considered as a binary $[nm, km]$ code. For some Reed-Solomon codes (as well as some generalizations), the weight distribution of the corresponding binary codes have been studied by Imamura et al. [86], Kasami and Lin [96], Kolev and Manev [114], Pradhan and Gupta [161], Retter [168], [169].

Codes with $d = n - k$ are called *almost MDS* codes. They have been discussed by de Boer [52], Dodunekov and Landgev [56], Faldum and Willems [58]. For these codes

$$A_i = (-1)^{i-n+k}\binom{k}{i-n+k}A_{n-k} + \binom{n}{i}\sum_{j=0}^{i-n+k-1}(-1)^j\binom{i}{j}\left(q^{i-j-n+k}-1\right)$$

for $n-k \leq i \leq n$. Some of these codes are good, others are not. For a discussion of the error detecting properties of almost MDS codes, see Kløve [111], [112].

3.10.5 Cyclic Codes

Several of the codes presented under other headings are cyclic. Here we will give some references to papers dealing with cyclic codes in general, and more particular, irreducible cyclic codes.

Barg and Dumer [29] gave two algorithms for computing the weight distribution of cyclic codes.

A cyclic code is called irreducible if the polynomial $h(z)$ in Theorem 2.1.2 is irreducible. An irreducible cyclic $[n,k;q]$ code C can be represented as follows.

$$C = \left\{\left(\mathrm{Tr}_1^k(x), \mathrm{Tr}_1^k(x\beta), \ldots, \mathrm{Tr}_1^k(x\beta^{n-1})\right) \,\middle|\, x \in GF(q^k)\right\},$$

where Tr is the trace function and β is a root of $h(z)$.

McEliece and Rumsey [146] and Helleseth et al. [80] gave general methods for computing the weight distribution of irreducible cyclic codes and also gave explicit expressions for several infinite classes of such codes. Their papers also contain a number of further references. Segal and Ward [175], [196] have computed the weight distribution of a number of irreducible cyclic codes.

One drawback with the use of cyclic codes is that they are not good for block synchronization. However, we can overcome this drawback by using some proper coset S of the cyclic code C since $P_{\mathrm{ue}}(C,p) = P_{\mathrm{ue}}(S,p)$.

3.10.6 Cyclic Redundancy Check (CRC) Codes

Consider a polynomial

$$g(z) = z^m + g_{m-1}z^{m-1} + \cdots + g_1 z + 1.$$

The $[n = k + m, k]$ code $C_{g,n}$ generated by $g(z)$ is a cyclic redundancy check (CRC) code. In general, the optimal choice of $g(z)$ will depend on k and p. We describe the case $m = 16$ in more detail since a number of such codes are used in practice. In 2.1 we referred a number of $[k + 16, k]$ CRC codes chosen as standards. In practice, a chosen $g(z)$ will be used for a range of k, and the best choice will depend on the criteria we use.

Methods for efficient computation of the weight distribution of CRC codes has been given by Fujiwara et al. [68], Miller et al. [151], Castagnoli et al. [46], Chun and Wolf [50]. For p sufficiently small, a code with larger minimum distance is better then one with smaller minimum distance. A candidate for an interesting generator polynomial is therefore a polynomial such that the minimum distance of the corresponding $[k+16, k]$ codes is large for a large range of values of k. Castagnoli et al. [45] considered the choice of $g(z)$ from this point of view. For example, the polynomial $z^{16}+z^{14}+z^{12}+z^{11}+z^9+z^8+z^7+z^4+z+1$ (used in IEC TC57) generates codes with minimum distance at least 6 for all lengths up to 151, and no other polynomial has this property. In Table 3.4 we list similar polynomials for other values of $d_{\min}$ (we write only the coefficients in the polynomial).

Based on an exhaustive search of all polynomial with $m = 16$, Castagnoli et al. [45] gave the optimal choices presented in Table 3.5. In the table, n_c is the largest length for which $d_{\min}(C_{g,n}) > 2$. A summary of the main properties of the codes is also given. For more details, we refer to [45].

$d_{\min} \geq$	coeff. of polynomial	length $\leq$
17	11111111111111111	17
12	10101101111101101	18
10	11101001000101111	21
9	11000111101010111	22
8	10001111110110111	31
7	10010011010110101	35
6	10011110101100101	151
5	10101100100110101	257
4	11010001011101011	32767

Table 3.4 Generator polynomial which generate codes of a given minimum distance to a given length.

Miller et al. [150], [151] considered the polynomials $g(z) = (1+z)p(z)$ where $p(z)$ is irreducible of degree 15. There are 896 such polynomials. They used the criterion that a polynomial $g(z)$ is better if the bound

$$P_{\text{ue}}(C_{g,n}, p) \leq 2^{-16}\left\{1 - 2(1-p)^n + (1-2p)^n\right\} \tag{3.20}$$

is satisfied over a larger range of values n. By this criterion the best polynomial turned out to be a polynomial that was optimal also by the criterion used by Castagnoli et al., namely the last polynomial in Table 3.5. For this polynomial, (3.20) is satisfied for all $n \leq n_c$.

coeff. of the polynomial g, properties of the codes $C_{g,n}$	n_c
10011110101100101 $d_{\text{min}} \geq 6$ for $n \leq n_c$, proper for $n \leq n_c$,	151
11111001010011111 $d_{\text{min}} \geq 6$ for $n \leq 130$, $d_{\text{min}} \geq 4$ for $n \leq n_c$, proper for $n \notin \{43,\ldots,48\} \cup \{189,\ldots,258\}$, $P_{\text{wc}}(C_{g,n}, 0, 0.5)/P_{\text{ue}}(C_{g,n}, 0.5) \leq 1.042$ for $n \leq n_c$	258
10101100100110101 $d_{\text{min}} \geq 5$ for $n \leq n_c$, proper for $n \leq n_c$,	257
10000000100011011 $d_{\text{min}} \geq 6$ for $n \leq 115$, $d_{\text{min}} \geq 4$ for $n \leq n_c$, proper when $n \leq 1127$ and $n \notin \{17,\ldots,28\} \cup \{31,\ldots,58\}$, $P_{\text{wc}}(C_{g,n}, 0, 0.5)/P_{\text{ue}}(C_{g,n}, 0.5) \leq 2.11$ for $n \leq n_c$	28658
11010001011101011 $d_{\text{min}} \geq 4$ for $n \leq n_c$, conjectured† to be proper for $n \leq n_c$, († Castagnoli et al. verified that $C_{g,n}$ is proper for $n \leq 256$ and a number of larger values of n.)	32767

Table 3.5 Generator polynomials and properties of the best *CRC* codes for $m = 16$.

Castagnoli et al. [46] and Wolf and Blakely [206] have done a similar analysis of polynomials $g(z)$ of degrees 24 and 32.

Wolf and Chun [207] considered an alternative model for channels with single bursts and the use of *CRC* codes to detect such burst. They also [50] describe special hardware to compute the probability of undetected error of *CRC* codes. Using this hardware they determined polynomials of the form $(z+1)p(z)$, where

$p(z)$ is irreducible, such that the corresponding shortened code is good for a large number of test lengths. They gave one polynomial for each degree from 8 through 39. These polynomials are listed in Table 3.6.

m = degree of $g(z)$	coeff. of the polynomial $p(z)$
8	10001001
9	101100011
10	1000101101
11	11100111001
13	1101110100111
14	10111010011001
15	110100001110111
16	1000011100101001
17	11000101110110111
18	100010010010111011
19	1011101000100010111
20	10010010111000010011
21	110101101110101100011
22	1000011100100110000101
23	10111101110110110100011
24	100010110000111010101011
25	1011100100110001111110101
26	11100101101000010110110001
27	100010110010010101110110111
28	1001001010010010101001111001
29	10001011001101101110011101001
30	100110110000010010001000101001
31	1110101100110110111010011111101
32	10100010100010001100010101011001
33	110111011100110101011110100001001
34	1001100010111001001000010010010011
35	10111111001000110111111010000101100001
36	110110010011100010101100101110110001
37	1001100001001000011100101110000101101
38	10010000110101010001110010111000000111
39	111011111010001100110101000111100111001

Table 3.6 Generator polynomials $g(z) = (z+1)p(z)$ generating CRC which are good for a range of shortened distances.

3.10.7 Z_4-linear Codes

Let, as usual, Z_4 denote the integers modulo 4; this is a ring. A linear code over Z_4 of length n is a module, that is, a subset C of Z_4^n such that if $\bar{u}, \bar{v} \in C$, then $a\bar{u} + b\bar{v} \in C$ for all $a, b \in Z_4$ (the arithmetic is done modulo 4). The dual code $C^\perp$ is defined the usual way via inner product (modulo 4). Let $\phi : Z_4 \to GF(2)^2$ be defined by

$$\phi(0) = (00), \quad \phi(1) = (01), \quad \phi(2) = (11), \quad \phi(3) = (10),$$

and $\phi : Z_4^n \to GF(2)^{2n}$ by

$$\phi(v_1, v_2, \ldots, v_n) = (\phi(v_1), \phi(v_2), \ldots, \phi(v_n)).$$

Finally, for a linear code C over Z_4 of length n we define the binary code $\phi(C)$ of length $2n$ by

$$\phi(C) = \{\phi(\bar{v}) \mid \bar{v} \in C\}.$$

Note that $\phi(C)$ is not in general linear; such codes have been termed Z_4-linear. For a more detailed description of these concepts we refer to the paper by Hammons et. al [76]. In particular they prove the following two results which are important in our context:

- A Z_4-linear code is distance invariant, in particular,

$$A_{\phi(C)}(z) = A^{\mathrm{w}}_{\phi(C)}(z),$$

- $A_{\phi(C^\perp)}(z) = A^{\mathrm{MW}}_{\phi(C)}(z).$

Note that both $\phi(C)$ and $\phi(C^\perp)$ may be non-linear and are not dual in the usual sense.

One class of codes which can be obtain this way is the *Kerdock codes* $\mathcal{K}(m)$ which are non-linear $(2^m, 2^{2m})$ codes. The weight distribution is

i	A_i
0	1
$2^{m-1} \pm 2^{(m-2)/2}$	$2^m\left(2^{m-1} - 1\right)$
2^{m-1}	$2^{m+1} - 2$
2^m	1

for even m (see [13, p.456] or [76]), and

i	A_i
0	1
$2^{m-1} \pm 2^{(m-1)/2}$	$2^{m-1}\left(2^{m-1}-1\right)$
2^{m-1}	$2^{m}\left(2^{m-1}+1\right)-2$
2^m	1

for odd m (see [76]).

The *Delsarte-Goethals* $\mathcal{DG}(m,d)$ *codes*, where $m \geq 4$ is even, are non-linear $(2^m, 2^{(m-1)(\lfloor m/2 \rfloor - d+1)+m+1})$ codes. In particular $\mathcal{DG}(m, m/2) = \mathcal{K}(m)$. The weight distribution of the $\mathcal{DG}(m, m/2-1)$ codes are

i	A_i
0	1
$2^{m-1} \pm 2^{m/2}$	$2^{m-2}\left(2^{m-1}-1\right)\left(2^m-1\right)/3$
$2^{m-1} \pm 2^{m/2+1}$	$2^{m}\left(2^{m-1}-1\right)\left(2^{m-1}+4\right)/3$
2^{m-1}	$2\left(2^m-1\right)\left(2^{2m-3}-2^{m-2}+1\right)$
2^m	1

(see [13, p.477]).

3.10.8 Some Other Classes of Codes

The Product of Two Single Parity Check Codes

Leung [131] determined $A_C(z)$ for the product of two binary single parity check codes. If C_1 and C_2 are $[l, l-1]$ and $[m, m-1]$ parity check codes respectively, then the weight distribution function of the product code is

$$\frac{1}{2^{l+m-1}} \sum_{i=0}^{m-1} \binom{m-1}{i} \left(z^i + z^{m-i}\right)^l .$$

The product code is proper for

$$\begin{aligned}(l,m) \in \; & \{(2,2),(2,3),(2,4),(2,5),(3,2),(3,3),(3,4),(3,5),\\ & (4,2),(4,3),(4,4),(5,2),(5,3)\},\end{aligned}$$

in all other cases the product code is not good.

Constant Weight Codes

Wang and Yang [194], [193] considered the binary $(n, 2\delta, m)$ codes, that is, the codes of binary vectors of length n, Hamming weight m, and minimum distance $\geq 2\delta$. Note that these codes are in general not linear. Wang and Yang showed that the codes are proper if $n \leq 4\delta$. The distance distribution of an $(n, 2, m)$ code is given by

$$A_i = \binom{m}{i}\binom{n-m}{i}.$$

Wang and Yang showed that that the $(5, 2, 2)$, $(5, 2, 3)$, $(6, 2, 3)$, $(7, 2, 3)$, $(7, 2, 4)$, and $(8, 2, 4)$ codes are proper. All other $(n, 2, m)$ codes with $n \geq 5$ are not good.

Melas Codes

With the notation introduced for BCH codes, the *Melas code* is the CRC code generated by the polynomial $\text{lcm}\{M_1(x), M_{-1}(x)\}$, where $M_i(x)$ denotes the minimal polynomial of α^i. For $q = 2$ the weight distribution was determined by Schoof and van der Vlugt [173] and for $q = 3$ by Van der Geer et. al. [190]. The weight distributions are expressed in terms of the trace of Hecke operators. The probability of undetected error for these codes has not been studied.

3.11 HOW TO FIND THE CODE YOU NEED

There are two main classes of problems we may encounter when we want to use an error detecting code on the BSC:

- how to calculate $P_{\text{ue}}(C, p)$ or $P_{\text{ue}}^{(t)}(C, p)$ for a given code,
- how to find an $[n, k]$ code for which $P_{\text{ue}}(C, p)$ or $P_{\text{ue}}^{(t)}(C, p)$ is below some given value.

In general these are difficult problems and a solution may be unfeasible. However, in many cases of practical interest, solutions may be found using the results and methods given in this chapter. Since the results are many and diverse, we give in this section a short "handbook" for how to attack these problems.

Given C and p, how to find or estimate $P_{\text{ue}}(C,p)$

The first thing to do is to check in section 3.10 if your code is in the list of codes with known weight distribution. If not, the weight distribution of the code or its dual may possibly be found by a complete listing of the code words (if k or $n-k$ are small, that is, not more than 30-35). Having found the weight distribution, $P_{\text{ue}}(C,p)$ can be computed from Theorem 2.3.1. If you have found the weight distribution of the dual code, $P_{\text{ue}}(C,p)$ can be found combining Theorems 2.3.1 and 3.4.1. Likewise, $P_{\text{ue}}^{(t)}(C,p)$ can be computed from (2.12) if the weight distribution is known, and from Theorem 2.2.2 and (2.12) if the weight distribution of the dual code is known.

If the weight distribution is not possible to obtain, you have to be satisfied with estimates for $P_{\text{ue}}(C,p)$. One upper bound is given in Theorem 3.5.10. Another is obtained by combining Theorems 2.2.11, 2.2.12 and 3.5.9. Any partial information you may have about the weight distribution may help to improve the bound thus obtained. If $p \le d/n$, Theorem 3.1.4 can be used. Lower bounds may be found using Theorems 3.5.1-3.5.6. Upper bounds on $P_{\text{ue}}^{(t)}(C,p)$ is obtained by combining (2.12) and Theorems 2.2.11 and 2.2.12.

Given C, a, and b, how to find or estimate $P_{\text{wc}}(C,a,b)$

To determine $P_{\text{wc}}(C,a,b)$ exactly is usually a harder problem than to determine $P_{\text{ue}}(C,p)$ for a particular p. In the special cases when we can prove that the code is proper, and $b \le 1/2$, then $P_{\text{wc}}(C,a,b) = P_{\text{ue}}(C,b)$. Likewise, if $b \le d/n$, then $P_{\text{wc}}(C,a,b) = P_{\text{ue}}(C,b)$ by Theorem 3.1.4. If the code is good, then by definition $P_{\text{wc}}(C,0,1/2) = P_{\text{ue}}(C,1/2) = (2^k-1)/2^n$.

With known weight distribution of the code, the algorithm based on Theorem 3.3.1 can be used to get as good an approximation to $P_{\text{wc}}(C,a,b)$ as we want. Otherwise, any upper bounds on the weight distribution (e.g. from Theorems 2.2.11 and 2.2.12) can be combined with Theorem 3.3.1 to give upper bounds on $P_{\text{wc}}(C,a,b)$. The last resort is to hope that $P_{\text{wc}}(C,a,b)$ is about average and that the upper bound on the average (3.14) is also an upper bound on $P_{\text{wc}}(C,a,b)$ (this is not very satisfactory, cfr. Theorem 3.3.4!).

Given p and a bound B, how to find an $[n, k]$ code C such that $P_{\text{ue}}(C, p) \le B$

It is possible that B is too small, that is, there are no $[n, k]$ codes C with the required property. The first thing to do is therefore to compare B with the lower bounds in Theorems 3.5.1-3.5.6. If B is not smaller than any these lower bounds, the next thing to do is to look through the list of codes in Section 3.10 to see if any of those satisfy the requirement. If not, possibly a code with the requirements can be obtained using the *-operation one or more times on any of these codes. If k or $n - k$ is not too large, the next possibility is to pick a number of $[n, k]$ codes at random and check them.

The problem of finding the $[n, k]$ code which minimizes $P_{\text{ue}}(C, p)$ is usually harder, but the line of attack would be the same. The solution for $k \le 4$ and for $n - k \le 4$ is given in Section 3.6. Further, if $n = s(2^k - 1)$, then the optimal code for all p is $C^{(s-1)*}$, where C is the $[2^k - 1, k]$ simplex code. Further, the $[2^m - 1, 2^m - 1 - m]$ Hamming codes are also optimal for all p.

Given a, b and a bound B, how to find an $[n, k]$ code C such that $P_{\text{we}}(C, a, b) \le B$

This problem should be attacked in the same way as the previous problem, the main difference is that now it is $P_{\text{we}}(C, a, b)$ which needs to be computed or estimated. As for a lower bound, note that $P_{\text{we}}(C, a, b) \ge (2^k - 1)/2^n$ if $a \le 1/2 \le b$.

We remark that for $[a, b] = [0, 1]$ an alternative line of attack is to use a stochastic transform, see pages 135ff.

How to choose a code which can be used for different sizes of the information, but with a fixed number of check bits

The problem can be stated as follows: find an $[m + r, r]$ code such that the shortened $[k + r, k]$ code is good for all k within some range, possibly for $1 \le k \le m$. The main choices of codes for this problem are CRC codes. You should consult the section on CRC codes (page 109) and, if need be, the references given there.

3.12 COMMENTS AND REFERENCES

3.1. Theorem 3.1.1 is due to Korzhik and Fink [10, page 184] (it was rediscovered by Padovani and Wolf [156]). A similar theorem for 1-good codes was given by Lin [137].

Theorem 3.1.2 is essentially due to Perry [158].

Theorem 3.1.3 was presented in Kløve [102].

Theorem 3.1.4 is an observation done by several people.

Sturm sequences are treated in many introductory books on numerical mathematics, e.g. Isaacson and Keller [8].

Theorem 3.1.5 is a special case of a method given in Kløve [101].

Theorems 3.1.6 and 3.1.7 are from Kasami et. al. [94].

3.2. Theorem 3.2.3 is due to Leontev [130].

The results on the variance are given here for the first time.

Theorem 3.2.4 is due to Korzhik [116].

Lemma 3.2.2 and Theorem 3.2.4 are due to Korzhik [116]. An upper bound on $\alpha(\bar{x})$ due to Lin [137] is identical to (3.4) for $d = 3$, weaker for $d > 3$.

3.3. The bounds on $P_{\text{wc}}(C, a, b)$ are from Kløve [107], [108]. Corollary 3.3.3 is due to Massey [145]. He also gave the bound $S_n \leq n$.

The sums $S_{n,k}$ are studied by Szpankowski [186]. In particular, (3.11) and (3.12) are due to him. The bounds (3.10) are from Kløve [107].

Theorem 3.3.4 is a special case of a theorem given in Kløve [99], the example to prove it given here is simpler. However, [99] gives the stronger result that there exist codes C with arbitrary large minimum distance and the property given in Theorem 3.3.4.

3.4. Theorem 3.4.1 was first given by Leontev [130] (for $q = 2$).

Theorem 3.4.2 has not been published before. It was presented in Kløve [102].

3.5. Theorem 3.5.1 is due to Korzhik [116].

Theorem 3.5.2 is due to Leontev [130].

Theorem 3.5.3 is due to Kløve [106]. A similar, but weaker bound was given by Wolf et al. [209].

Lemma 3.5.1 and Theorem 3.5.4 are due to Levenshtein [135].

Theorems 3.5.6 and 3.5.7 are due to Levenshtein [136].

Theorems 3.5.9 and 3.5.10 are new. A weaker upper bound based on the same idea was given in Kasami et al. [94].

Theorem 3.5.11 is due to Leontev [130].

Theorems 3.5.12 and 3.5.13 are due to Levenshtein [135]. A more general bound, which has the bounds in Theorems 3.5.11 and 3.5.12 as special cases, was given by Katsman [97].

Theorems 3.5.14, 3.5.15, and 3.5.16 are all due to Levenshtein [135] and [136].

3.6. Theorem 3.6.2 is a special case of a theorem given by Kløve [103].

3.7. The results and examples presented are mostly new. However, other examples to illustrate some of these results have been given by Leung et.al. [131], [132], [133].

3.8. The discussion of $Q_t(C,p)$ is taken from Kløve [99].

For more details on the concatenated codes with $m = 1$ we refer to Kløve and Miller [113].

Bounds on the probability of undetected error for a concatenated code with Reed-Solomon outer code are given in Korzhik and Fink [10, pages 188-190].

For the concatenated codes with $m > 1$, see Kasami et al. [93] which have a further discussion and some examples.

Many systems use a concatenation of a convolutional code for error correction and a block code (usually a CRC code) for error detection. Some papers discussing various such systems are [77], [176], [212].

3.9. The observation that the problem of computing $P_{\mathrm{ue}}(C,p)$ in general is an $\mathcal{NP}$ hard problem seems to be new.

4

CODES FOR OTHER CHANNELS

4.1 THE Z-CHANNEL

We showed in (2.11) that for any code C we have

$$P_{\text{ue}}(C, Z_p) = \frac{1}{\#C} \sum_{\bar{x} \in C} \sum_{\substack{\bar{y} \in C \\ \bar{y} < \bar{x}}} p^{w_{\text{H}}(\bar{x}) - w_{\text{H}}(\bar{y})} (1-p)^{w_{\text{H}}(\bar{y})}. \tag{4.1}$$

A code C is called *perfect for error detection* on the Z-channel if $P_{\text{ue}}(C, Z_p) = 0$ for all p. By (4.1), C is perfect if and only if $\bar{y} \not< \bar{x}$ for all $\bar{y}, \bar{x} \in C$, $\bar{y} \neq \bar{x}$. The largest perfect C of length n is obtained by taking all vectors of weight $\lfloor \frac{n}{2} \rfloor$.

A perfect systematic $\left(k + \lceil \log_2 k \rceil, 2^k\right)$ code can be constructed as follows:

$$C_k = \{(\bar{x} | r(\bar{x})) \mid \bar{x} \in GF(2)^k\}$$

where $r(\bar{x}) \in GF(2)^{\lceil \log_2 k \rceil}$ is obtained by first taking the binary expansion of $w_{\text{H}}(\bar{x})$ and then taking the binary complement (changing 0 to 1 and 1 to 0). E.g.

$$C_3 = \{00011, 00110, 01010, 01101, 10010, 10101, 11001, 11100\}.$$

We note that C_k has 2^k code words whereas the non-systematic perfect code of the same length given above has approximately $2^k \sqrt{2k/\pi}$ code words.

A general construction of (non-perfect) error detecting codes for the Z-channel is given in the next theorem.

Theorem 4.1.1 *Let a_i be integers for $1 \le i \le k$ such that*

$$0 \le a_1 < a_2 < \cdots < a_k \le n.$$

Let

$$C = \{\bar{x} \in GF(2)^n \mid w_{\mathrm{H}}(\bar{x}) = a_i \text{ for some } i\}.$$

Then

$$P_{\mathrm{ue}}(C, Z_p) = \frac{1}{\sum_{i=1}^{k} \binom{n}{a_i}} \sum_{i=2}^{k} \sum_{j=1}^{i-1} \binom{n}{a_i} \binom{a_i}{a_j} p^{a_i - a_j} (1-p)^{a_j}.$$

Proof: There are $\binom{n}{a_i}$ vectors $\bar{x} \in C$ of Hamming weight a_i. For each of these there are $\binom{a_i}{a_j}$ vectors $\bar{y} \in C$ of Hamming weight a_j such that $\bar{y} < \bar{x}$.

Example. Let $a_i = 2i - 1$ for $i = 1, 2, \ldots, k$ where $k = \lceil n/2 \rceil$, that is, we take all vectors of odd weight. The corresponding code C_{o} detects all single asymmetric errors (and many others), and from Theorem 4.1.1 we get after simplifications:

$$P_{\mathrm{ue}}(C_{\mathrm{o}}, Z_p) = \frac{1}{2} + \frac{1}{2}(1-p)^n - \left(1 - \frac{p}{2}\right)^n - \left(\frac{1}{2} - \frac{1}{2^n}\right) p^n.$$

Another possibility is to let $a_i = 2i$ for $i = 0, 1, \ldots, k$ where $k = \lfloor n/2 \rfloor$, that is, we take all vectors of even weight. For the corresponding code C_{e} we get

$$P_{\mathrm{ue}}(C_{\mathrm{e}}, Z_p) = \frac{1}{2} + \frac{1}{2}(1-p)^n - \left(1 - \frac{p}{2}\right)^n + \left(\frac{1}{2} - \frac{1}{2^n}\right) p^n.$$

In particular, $\#C_{\mathrm{o}} = \#C_{\mathrm{e}}$ and $P_{\mathrm{ue}}(C_{\mathrm{o}}, Z_p) < P_{\mathrm{ue}}(C_{\mathrm{e}}, Z_p)$ for all $n > 1$, $p \neq 0$.

4.2 CHANNELS WITH MEMORY

We will restrict our consideration of channels with memory to models with additive noise. Let K be some channel with additive noise. Let $P(\bar{x})$ be the probability of an error vector $\bar{x} \in GF(q)^n$. For any $\bar{e} \in GF(2)^n$, let

$$E(\bar{e}) = \{\bar{x} \in GF(q)^n \mid \chi(\bar{x}) = \chi(\bar{e})\}.$$

By definition of an additive channel we have $P(\bar{x}) = P(\bar{e})$ for all $\bar{x} \in E(\bar{e})$.

Let C be a linear $[n,k;q]$ code. Then according to the definition of such models (see Chapter 1) and the principles of error detection (see Section 2.3) we get

$$P_{\text{ue}}(C,K) = \sum_{\bar{x}\in C\setminus\{\bar{0}\}} P(\bar{x}). \tag{4.2}$$

To calculate $P_{\text{ue}}(C,K)$ by (4.2) we could use the different models for the channels described in Chapter 1, but it is usually very difficult to compute $P(\bar{e})$ for all $\bar{e}$ for powerful codes. Therefore we will give some simpler formulas and bounds on $P_{\text{ue}}(C,K)$, at least in some cases. In particular, we will show that P_{ue} can easily be calculated for the stochastic channel transforms corresponding to some degenerated channels.

General Bounds on $P_{\text{ue}}(C,K)$

We first show that Theorem 3.2.1 generalizes immediately. Let $\mathcal{C}$ be a set of $[n,k;q]$ codes. For $\bar{x}\in GF(q)^n$, let

$$\alpha(\bar{x}) = \alpha_{\mathcal{C}}(\bar{x}) = \frac{\#\{C\in\mathcal{C} \mid \bar{x}\in C\}}{\#\mathcal{C}}.$$

Further, let

$$P_{\text{ef}}(n) = P(\bar{0}),$$

the probability of an error free block of length n.

Theorem 4.2.1 *Let $\mathcal{C}$ be a set of $[n,k;q]$ codes and K any channel with additive noise. Then*

$$E(P_{\text{ue}}(C,K)) = \sum_{\bar{x}\in GF(q)^n\setminus\{\bar{0}\}} \alpha(\bar{x})P(\bar{x}).$$

Proof: We have

$$\begin{aligned}
E(P_{\text{ue}}(C,K)) &= \frac{1}{\#\mathcal{C}}\sum_{C\in\mathcal{C}} P_{\text{ue}}(C,K) \\
&= \frac{1}{\#\mathcal{C}}\sum_{C\in\mathcal{C}}\sum_{\bar{x}\in C\setminus\{\bar{0}\}} P(\bar{x}) \\
&= \sum_{\bar{x}\in GF(q)^n\setminus\{\bar{0}\}} P(\bar{x})\frac{1}{\#\mathcal{C}}\sum_{\bar{x}\in C\in\mathcal{C}} 1.
\end{aligned}$$

Corollary 4.2.1 *Let C be the set of all $[n, k; q]$ codes and K any channel with additive noise. Then*

$$E(P_{\text{ue}}(C, K)) = \frac{q^k - 1}{q^n - 1}(1 - P_{\text{ef}}(n)). \tag{4.3}$$

In the same way we get $E(P_{\text{ue}}(C, K)) = q^{k-n}(1 - P_{\text{ef}}(n))$ for the set of all systematic $[n, k; q]$ codes.

Corollary 4.2.1 is important, because even if we do not have any information about the probability distribution of the error patterns on some channel K, it is still possible to say that there exists some $[n, k; q]$ code C such that

$$P_{\text{ue}}(C, K) \leq \frac{1}{q^{n-k}}. \tag{4.4}$$

Unfortunately, there is no known analogue to Theorem 3.2.4 in the general case.

Worst-Case Probability of Error on the Local Binomial Channel

We will now take a closer look at the special case of the LBC-1 channels. In analogy to what we did for the BSC, we define

$$P_{\text{wc}}(C, a, b) = \max\left\{P_{\text{ue}}(p_1, p_2, \ldots, p_n) \;\middle|\; p_i \in [a, b] \text{ for } 1 \leq i \leq n\right\}.$$

Theorem 4.2.2 *Let C be an $[n, k; q]$ code, and let $0 \leq a \leq b \leq 1$. For $X \subseteq \{1, 2, \ldots, n\}$, define $\bar{z}(X) = \bar{z}$ by*

$$z_i = \begin{cases} b & \text{if } i \in X, \\ a & \text{if } i \notin X. \end{cases}$$

Then

$$P_{\text{wc}}(C, a, b) = \max\left\{P_{\text{ue}}\left(C, \bar{z}(X)\right) \;\middle|\; X \subseteq \{1, 2, \ldots, n\}\right\}.$$

Proof: For convenience, we write

$$Q = \{\bar{p} \mid a \leq p_i \leq b \text{ for } 1 \leq i \leq n\}.$$

We prove by induction on j that there exists a vector $\bar{y} \in Q$ such that

$$y_i \in \{a, b\} \quad \text{for} \quad 1 \le i \le j, \tag{4.5}$$
$$P_{\text{ue}}(C, \bar{p}) \le P_{\text{ue}}(C, \bar{y}) \quad \text{for all} \quad \bar{p} \in Q. \tag{4.6}$$

First, let $j = 0$. Since Q is a closed set and $P_{\text{ue}}(C, \bar{p})$ is a continuous function of $\bar{p}$ on Q , it obtains its maximum on Q, that is, there exists a $\bar{y} \in Q$ such that (4.6). Moreover, (4.5) is trivially true for $j = 0$. This proves the induction basis.

For the induction step, let $j > 0$ and suppose that the statement is true for $j - 1$; let $\bar{y} \in Q$ such that $y_i \in \{a, b\}$ for $1 \le i \le j - 1$ and (4.6) is satisfied. For a $\bar{c} \in C$ we have

$$\prod_{i=1}^{n} \left(\frac{y_i}{q-1}\right)^{w_{\text{H}}(c_i)} (1 - y_i)^{1-w_{\text{H}}(c_i)}$$
$$= \begin{cases} y_j \frac{1}{q-1} \prod_{i \ne j} \left(\frac{y_i}{q-1}\right)^{w_{\text{H}}(c_i)} (1 - y_i)^{1-w_{\text{H}}(c_i)} & \text{if } c_j \ne 0, \\ (1 - y_j) \prod_{i \ne j} \left(\frac{y_i}{q-1}\right)^{w_{\text{H}}(c_i)} (1 - y_i)^{1-w_{\text{H}}(c_i)} & \text{if } c_j = 0. \end{cases}$$

Hence

$$P_{\text{ue}}(C, \bar{y}) = \alpha_j y_j + \beta_j$$

where

$$\begin{aligned} \alpha_j &= \frac{1}{q-1} \sum_{\substack{\bar{c} \in C \\ c_j \ne 0}} \prod_{i \ne j} \left(\frac{y_i}{q-1}\right)^{w_{\text{H}}(c_i)} (1 - y_i)^{1-w_{\text{H}}(c_i)} \\ &\quad - \sum_{\substack{\bar{c} \in C \setminus \{\bar{0}\} \\ c_j = 0}} \prod_{i \ne j} \left(\frac{y_i}{q-1}\right)^{w_{\text{H}}(c_i)} (1 - y_i)^{1-w_{\text{H}}(c_i)} \\ \beta_j &= \sum_{\substack{\bar{c} \in C \setminus \{\bar{0}\} \\ c_j = 0}} \prod_{i \ne j} \left(\frac{y_i}{q-1}\right)^{w_{\text{H}}(c_i)} (1 - y_i)^{1-w_{\text{H}}(c_i)}. \end{aligned}$$

If $\alpha_j > 0$ we must have $y_j = b$ since $\bar{y}$ is a maximum point, and so (4.5) is true. Similarly, if $\alpha_j < 0$ we must have $y_j = a$ and again (4.5) is true. Finally, if $\alpha_j = 0$, then $P_{\text{ue}}(C, \bar{y})$ is independent of the value of y_j. Therefore, we may replace y_j by any other value in $[a, b]$, e.g. a which makes (4.5) true also in this case. This proves the statement for j and the induction is complete. For $j = n$ we see that (4.5) and (4.6) implies that there exists an X such that $P_{\text{wc}}(C, a, b) = P_{\text{ue}}\Big(C, \bar{z}(X)\Big)$.

Lemma 4.2.1 *We have*

$$P_{\text{ue}}\Big(C, \bar{z}(X)\Big) = \frac{|\{\bar{c} \in C \mid \chi(\bar{c}) = X\}|}{(q-1)^{|X|}}.$$

Proof: In the case we consider, $\bar{z}(X)$ is given by

$$z_i = \begin{cases} 1 & \text{if } i \in X, \\ 0 & \text{if } i \notin X. \end{cases}$$

and so

$$\left(\frac{z_i}{q-1}\right)^{w_{\text{H}}(c_i)} (1-z_i)^{1-w_{\text{H}}(c_i)} = \begin{cases} 1 & \text{if } w_{\text{H}}(c_i) = p = 0, \\ 1/(q-1) & \text{if } w_{\text{H}}(c_i) = p = 1, \\ 0 & \text{otherwise.} \end{cases}$$

Hence

$$\prod_{i=1}^{n} \left(\frac{z_i}{q-1}\right)^{w_{\text{H}}(c_i)} (1-z_i)^{1-w_{\text{H}}(c_i)} = \begin{cases} 1/(q-1)^{|X|} & \text{if } \chi(\bar{c}) = X, \\ 0 & \text{otherwise.} \end{cases}$$

Summing over all $\bar{c} \in C \setminus \{\bar{0}\}$, the lemma follows.

Theorem 4.2.3 *Let C be an $[n, k, d; q]$ code. Then*

$$P_{\text{wc}}(C, 0, 1) = \frac{1}{(q-1)^{d-1}}.$$

Proof: First, let $\bar{c}$ be a code word in C of minimum weight d, and let $X = \chi(\bar{c})$. Then $\chi(a\bar{c}) = X$ for all $a \neq 0$. Hence

$$P_{\text{wc}}(C, 0, 1) \geq P_{\text{ue}}\Big(C, \bar{z}(X)\Big) \geq \frac{q-1}{(q-1)^d} = \frac{1}{(q-1)^{d-1}}. \tag{4.7}$$

On the other hand, let X be the support of some non-zero code word. Let

$$B_X = \{\bar{c}_X \mid \bar{c} \in C \text{ and } \chi(\bar{c}) = X\}$$

where $\bar{c}_X$ denotes the vector of length $|X|$ obtained by puncturing all positions of $\bar{c}$ not in X (the elements in the punctured positions are all zero). Let U_X be the vector space generated by B_X. Let r be the dimension and δ the minimum distance of U_X. Clearly, $d \leq \delta$. By the Singleton bound we have $\delta \leq |X| - r + 1$. Hence $r \leq |X| - (d-1)$. By Theorem 2.2.12, U_X contains not more than

$(q-1)^{|X|-(d-1)}$ code words of weight $|X|$. In particular, $|B_X| \le (q-1)^{|X|-(d-1)}$. By Lemma 4.2.1 we have

$$P_{\text{ue}}\Big(C, \bar{z}(X)\Big) = \frac{|B_X|}{(q-1)^{|X|}} \le \frac{(q-1)^{|X|-(d-1)}}{(q-1)^{|X|}} = \frac{1}{(q-1)^{d-1}}.$$

By Theorem 4.2.2 we have

$$P_{\text{wc}}(C,0,1) = \max\Big\{P_{\text{ue}}\Big(C, \bar{z}(X)\Big) \;\Big|\; X \subseteq \{1,2,\ldots,n\}\Big\} \le \frac{1}{(q-1)^{d-1}}.$$

This, together with (4.7), proves the theorem.

Lemma 4.2.2 *Consider the interval* $[0,(q-1)/q]$. *We have*

$$P_{\text{ue}}\Big(C, \bar{z}(X)\Big) = \frac{|\{\bar{c} \in C \setminus \{\bar{0}\} \mid \chi(\bar{c}) \subseteq X\}|}{q^{|X|}}.$$

Proof: First we note that if $i \in X$, then $z_i = (q-1)/q$, and so

$$\frac{z_i}{q-1} = 1 - z_i = \frac{1}{q}.$$

Therefore

$$\left(\frac{z_i}{q-1}\right)^{w_{\text{H}}(c_i)} (1-z_i)^{1-w_{\text{H}}(c_i)} = \frac{1}{q}$$

for any value of c_i. If $i \notin X$, then $z_i = 0$ and so

$$\left(\frac{z_i}{q-1}\right)^{w_{\text{H}}(c_i)} (1-z_i)^{1-w_{\text{H}}(c_i)} = \begin{cases} 1 & \text{if } c_i = 0, \\ 0 & \text{if } c_i \neq 0. \end{cases}$$

Hence

$$\prod_{i=1}^{n} \left(\frac{z_i}{q-1}\right)^{w_{\text{H}}(c_i)} (1-z_i)^{1-w_{\text{H}}(c_i)} = \begin{cases} q^{-|X|} & \text{if } \chi(\bar{c}) \subseteq X, \\ 0 & \text{if } \chi(\bar{c}) \not\subseteq X. \end{cases}$$

Summing over all non-zero $\bar{c}$ in C, the lemma follows.

For $X \subseteq \{1,2,\ldots,n\}$, let

$$V_X = \{\bar{c} \in C \mid \chi(\bar{c}) \subseteq X\}.$$

We note that V_X is a vector space, i.e. V_X is a subcode of C. If the dimension of V_X is r, then

$$\left|\left\{\bar{c} \in C \setminus \{\bar{0}\} \mid \chi(\bar{c}) \subseteq X\right\}\right| = q^r - 1$$

and

$$d_r \leq \left|\chi(V_X)\right| \leq |X|.$$

Hence

$$P_{\text{ue}}\left(C, \bar{z}(X)\right) \leq \frac{q^r - 1}{q^{d_r}}. \tag{4.8}$$

Moreover, for each r, $1 \leq r \leq k$, there exists a set X such that $|X| = d_r$ and $\dim(V_X) = r$. Combining this with Theorem 4.2.2 and Lemma 4.2.2 we get the following result.

Theorem 4.2.4 *For an $[n, k; q]$ code C we have*

$$P_{\text{wc}}(C, 0, (q-1)/q) = \max_{1 \leq r \leq k} \left\{\frac{q^r - 1}{q^{d_r}}\right\}.$$

Next, we state a lemma whose simple proof is omitted.

Lemma 4.2.3 *Let l, m, and t be positive integers. Then*

$$\text{a)} \quad \frac{q^l - 1}{q^m} < \frac{q^{l+t} - 1}{q^{m+t}}, \tag{4.9}$$

$$\text{b)} \quad \frac{q^l - 1}{q^m} > \frac{q^{l+t} - 1}{q^{m+u}} \quad \textit{for } u \geq t + 1. \tag{4.10}$$

We can now restate Theorem 4.2.4.

Theorem 4.2.5 *Let C be an $[n, k, d; q]$ code. Let*

$$s = \max\left\{r \,\middle|\, 1 \leq r \leq k \text{ and } d_r = d_1 + (r - 1)\right\}.$$

Then

$$P_{\text{wc}}(C, 0, (q-1)/q) = \frac{q^s - 1}{q^{d+s-1}}.$$

Proof: First, since $d_s - d_r = s - r$ for $1 \le r \le s$, we have, by Lemma 4.2.3 a),

$$\frac{q^r - 1}{q^{d_r}} < \frac{q^s - 1}{q^{d_s}}$$

for $1 \le r < s$. For $s < r \le k$ we have, by (2.2), that $d_r - d_s \ge r - s + 1$, and by Lemma 4.2.3 b) we get

$$\frac{q^s - 1}{q^{d_s}} > \frac{q^r - 1}{q^{d_r}}.$$

Hence, by Theorem 4.2.4, we get

$$P_{\text{wc}}(C, (q-1)/q) = \frac{q^s - 1}{q^{d_s}} = \frac{q^s - 1}{q^{d+s-1}}.$$

We give some corollaries of Theorem 4.2.5. Usually only part of the weight hierarchy is needed to determine $P_{\text{wc}}(C, 0, (q-1)/q)$, sometimes only d_1. For instance, by (2.1) we have $d_2 > d_1 + 1$ if $d_1 \ge q+1$. Hence we have the following corollary.

Corollary 4.2.2 *Let C be an $[n, k, d; q]$ code with minimum distance $d > q$. Then*

$$P_{\text{wc}}(C, 0, (q-1)/q) = \frac{q-1}{q^d}.$$

From Theorem 4.2.4 we see that

$$P_{\text{wc}}(C, 0, (q-1)/q) \ge \frac{q^k - 1}{q^{d_k}} \ge \frac{q^k - 1}{q^n}.$$

In particular, $P_{\text{wc}}(C, (q-1)/q) = (q^k - 1)/q^n$ if and only if $n = d_k = d_1 + k - 1$, that is, if and only if the code is MDS.

Corollary 4.2.3 *For an $[n, k; q]$ code C we have*

$$P_{\text{wc}}(C, 0, (q-1)/q) \ge \frac{q^k - 1}{q^n}$$

with equality if and only if C is an MDS code.

Remark. The average value of $P_{ue}(C, p_1, p_2, \ldots, p_n)$ over all $[n, k; q]$ codes can be shown to be

$$E\Big(P_{ue}(C, p_1, p_2, \ldots, p_n)\Big) = \frac{q^k - 1}{q^n - 1}\left\{1 - \prod_{i=1}^{n}(1 - p_i)\right\},$$

and so

$$\max\left\{E\Big(P_{ue}(C, p_1, p_2, \ldots, p_n)\Big) \,\Big|\, 0 \le p_i \le (q-1)/q\right\} = \frac{q^k - 1}{q^n}.$$

In analogy to what we did for the BSC, we could call a code good if $P_{wc}(C, 0, (q-1)/q) = (q^k - 1)/q^n$. However, we note that this is a very strong condition, which by Corollary 4.2.3 is satisfied only by MDS codes.

It is often difficult to determine the weight hierarchy. Therefore it is useful to have good bounds. Corollary 4.2.3 gives such a lower bound. Theorem 4.2.5 and Lemma 4.2.3 a) give the following general bounds.

Corollary 4.2.4 *Let C be an $[n, k, d; q]$ code. Then*

$$\frac{q-1}{q^d} \le P_{wc}(C, 0, (q-1)/q) \le \frac{q^k - 1}{q^{d+k-1}} = \frac{q - q^{-(k-1)}}{q^d}.$$

Bounds on $P_{ue}(C, K)$ on Some Particular Channels

Let

$$p(n, i) = \max\{p(\bar{e}) \mid \bar{e} \in GF(2)^n \text{ and } w_H(\bar{e}) = i\}, \tag{4.11}$$

where $p(\bar{e})$ is the error pattern probability function for some given channel K. Then we have the following upper bound:

$$P_{ue}(C, K) \le \sum_{i=d}^{n} A_i(C) p(n, i). \tag{4.12}$$

We can use all methods described in Chapters 2 and 3 to calculate or estimate the values $A_i(C)$. A cruder, but simpler, bound can be obtained for any $[n, k, d; q]$ code C used for combined error correction and detection:

$$P_{ue}^{(t)}(C, K) \le \sum_{m=d-t}^{n} \tilde{p}(n, m), \tag{4.13}$$

where $\tilde{p}(n,m)$ is the probability that there are m errors in a block of n symbols, for the given channel K. This follows from the fact that the set of undetectable error patterns is a subset of the set of error patterns of Hamming weight at least $d-t$.

From the results in 3.8 we can obtain similarly a bound on the corresponding $P_{\text{ue}}^{(t)}(C,K)$ for any binary channel with additive noise:

$$P_{\text{ue}}^{(t)}(C,K) \le \sum_{i=d}^{n} A_i(C) \sum_{j=0}^{n} N_t(i,j)p(n,j), \tag{4.14}$$

where $N_t(i,j)$ is given in Theorem 2.2.10 and $p(n,j)$ is given by (4.11).

Another version of the upper bound in terms of the coset weight distribution is the following:

$$P_{\text{ue}}^{(t)}(C,K) \le \sum_{i=0}^{N} \sum_{j=1}^{n} N_{ij} p(n,j) - \sum_{m=1}^{t} \tilde{p}(n,m), \tag{4.15}$$

where

- $N = \sum_{i=1}^{t} \binom{n}{i}$,
- N_{ij} is the number of words of weight j in the i-th coset,
- $p(n,j)$ is given by (4.11),
- $\tilde{p}(n,m)$ has the same meaning as in (4.13).

For the composite channel (CC) described in Chapter 1 we get results similar to the BSC:

$$P_{\text{ue}}(C,CC) = \sum_{i=d}^{n} A_i(C) \int_{\mu} p(\mu)^i (1-p(\mu))^{n-i} w(\mu) d\mu, \tag{4.16}$$

where C is an $[n,k,d]$ code; further $p(\mu)$ and $w(\mu)$ where defined in Chapter 1.

All formulas and bounds on $A_i(C)$ given in Chapter 3 can be used to calculate or estimate $P_{\text{ue}}(C,CC)$ when combined with (4.16).

For the RFC model we get from (1.20), (1.21), and (4.12) that

$$P_{\text{ue}}(C,RFC) \le \sum_{i=d}^{n} \frac{A_i(C)}{(2\alpha h^2)^i (1-r^2)^{i-1}}. \tag{4.17}$$

For the FPMC model we get from (4.12)

$$P_{\text{ue}}(C, FPMC) \leq \sum_{i=d}^{n} A_i(C) \max\{\bar{a}\mathbf{A}_{e_1}\mathbf{A}_{e_2}\cdots\mathbf{A}_{e_n}\bar{b}^{\text{t}}\}, \tag{4.18}$$

where the maximum is taken over all binary vectors $(e_1, e_2, \ldots, e_n)$ of weight i. In particular, if $(1, 1, \ldots, 1, 0, 0, \ldots, 0)$ gives the maximum for each i, then

$$P_{\text{ue}}(C, FPMC) \leq \sum_{i=d}^{n} A_i(C)\bar{a}\mathbf{A}_1^i\mathbf{A}_0^{n-i}\bar{b}^{\text{t}}. \tag{4.19}$$

For the LBC, the following simple bound holds:

$$P_{\text{ue}}(C, LBC) \leq \sum_{i=d}^{n} A_i(C)\left(\max\mathcal{P}\right)^i\left(1-\min\mathcal{P}\right)^{n-i}, \tag{4.20}$$

where $\mathcal{P}$ is the set of symbol probabilities for the LBC. Obviously, if for the binary LBC $\mathcal{P}$ is the full achievable interval $\left[0, \frac{1}{2}\right]$, then we get a trivial bound.

An interesting model close to the LBC is obtained when $p_1 = p_2 = \cdots = p_k = p_I$ and $p_{k+1} = p_{k+2} = \cdots = p_n = p_C$, that is, all the information symbols have probability p_I and all the check symbols have probability p_C. Such a situation is typical for a system with information feedback. It is in general very difficult to calculate $P_{\text{ue}}(C)$ for such a model; it requires more knowledge about the code than its weight distribution, cfr. Theorem 2.2.4. In the particular case when $p_C = 0$ (noiseless return channel) we get

$$P_{\text{ue}}(C) = \sum_{i=d(C')}^{k} A_i(C')p_I^i(1-p_I)^{k-i}, \tag{4.21}$$

where C' is the subcode of C having all check symbols zero.

A very simple precise formula for P_{ue} can be derived for the channel model with a burst of errors (BEC), described in Chapter 1, when we use cyclic codes. In fact, it is known that for any binary cyclic $[n, k]$ code C, the number $N_i(C)$ of code words coinciding with bursts of length i are given by (see e.g. [207]):

$$N_i(C) = \begin{cases} 0 & \text{if } 1 \leq n-k, \\ k & \text{if } i = n-k+1, \\ 2^{i-(n-k-2)}(n-i+1) & \text{if } i \geq n-k+2. \end{cases} \tag{4.22}$$

From the definition of the BEC model we get

$$P_{\text{ue}}(C, BEC) = f(n-k+1, n)k + \sum_{i=n-k+2}^{n} \frac{f(i,n)(n-i+1)}{2^{n-k-2-i}}, \tag{4.23}$$

where $f(i,n)$ is the function which determines the BEC model.

Bounds for LBC With Concatenated Codes

Consider a concatenated code C with inner $[n_1, k_1, d_1]$ code C_1 and outer $[n_2, k_2; q]$ shortened RS code C_2. We restrict our attention to the following decoding algorithms: the inner code corrects at most $t \leq \lfloor \frac{d_1}{2} \rfloor - 1$ errors and detects further errors, the outer code corrects at most $n_2 - k_2 - l$ erasures (blocks where the inner code detected errors), where $0 \leq l \leq n_2 - k_2$, and further detects errors.

Theorem 4.2.6 *Let C be the concatenated code with the decoding algorithm described above. Let $s = n_1 m$ for some positive integer m. Then*

$$P_{\text{ue}}(C, LBC\text{-}s) \leq \binom{n_2}{l}\left[\left(\frac{d_1 - t}{n_1}\right)^{d_1 - t}\left(1 - \frac{d_1 - t}{n_1}\right)^{n_1 - d_1 + t}\sum_{j=0}^{t}\binom{n_1}{j}\right]^l .$$

Proof: It is obvious that an error will be undetected by the the outer code if and only if all the unerased positions coincide with the corresponding positions of a code word different from the transmitted code word. Let $i_1, i_2, \ldots, i_s$, where $s \geq k_2 + l$, be the non-erased positions. Then

$$P_{\text{ue}}(C, LBC\text{-}s) = \sum_{(\hat{a}_{i_1}, \hat{a}_{i_2}, \ldots, \hat{a}_{i_s}) \in A} \prod_{j=1}^{s} P(\hat{a}_{i_j} | a_{i_j}), \tag{4.24}$$

where $P(\hat{a}_{i_j} | a_{i_j})$ is the transition probabilities for the symbols in the i_j-th position, and A is the set of all tuples $(\hat{a}_{i_1}, \hat{a}_{i_2}, \ldots, \hat{a}_{i_s})$ occuring in the outer code, except the tuple $(a_{i_1}, a_{i_2}, \ldots, a_{i_s})$ in the transmitted code word. For each subset $\{\tau_1, \tau_2, \ldots, \tau_l\}$ of $\{i_1, i_2, \ldots, i_s\}$, let $A_{\tau_1, \tau_2, \ldots, \tau_l}$ denote the set of tuples $(\hat{a}_{i_1}, \hat{a}_{i_2}, \ldots, \hat{a}_{i_s}) \in A$ such that $(\hat{a}_{\tau_1}, \hat{a}_{\tau_2}, \ldots, \hat{a}_{\tau_l}) \neq (a_{\tau_1}, a_{\tau_2}, \ldots, a_{\tau_l})$. Clearly $A \subseteq \bigcup_{\{\tau_1, \tau_2, \ldots, \tau_l\} \subseteq \{i_1, i_2, \ldots, i_s\}} A_{\tau_1, \tau_2, \ldots, \tau_l}$ and therefore

$$\begin{aligned} P_{\text{ue}}(C, LBC\text{-}s) \leq & \sum_{\{\tau_1, \tau_2, \ldots, \tau_l\} \subseteq \{i_1, i_2, \ldots, i_s\}} \sum_{\substack{(\hat{a}_{\tau_1}, \hat{a}_{\tau_2}, \ldots, \hat{a}_{\tau_l}) \\ \neq (a_{\tau_1}, a_{\tau_2}, \ldots, a_{\tau_l})}} \prod_{r=1}^{l} P(\hat{a}_{\tau_r} | a_{\tau_r}) \\ & \sum_{\substack{(\hat{a}_{i_1}, \hat{a}_{i_2}, \ldots, \hat{a}_{i_s}) \in A \\ i_j \neq \tau_r}} \prod_{\substack{j=1 \\ i_j \neq \tau_r}}^{s} P(\hat{a}_{i_j} | a_{i_j}). \end{aligned}$$

Since the inner code has minimum distance d_1 and corrects at most t errors, we have

$$P(\hat{a}_{\tau_r}|a_{\tau_r}) \leq \left(p^{(\tau_r)}\right)^{d_1-t}\left(1-p^{(\tau_r)}\right)^{n-d_1+t}, \tag{4.25}$$

where $p^{(\tau_r)}$ is the (binary) symbol probability on the τ_r-th block of the inner code. Substituting (4.25) into the previous equation we get

$$P_{\text{ue}}(C, LBC\text{-}s) \leq \left\{\max_{p^{(\tau_r)}}\left\{\left(p^{(\tau_r)}\right)^{d_1-t}\left(1-p^{(\tau_r)}\right)^{n-d_1+t}\right\}\right\}$$

$$\sum_{\substack{\{\tau_1,\tau_2,\ldots,\tau_l\}\\ \subseteq\{i_1,i_2,\ldots,i_s\}}}\ \sum_{\substack{(\hat{a}_{\tau_1},\hat{a}_{\tau_2},\ldots,\hat{a}_{\tau_l})\\ \neq(a_{\tau_1},a_{\tau_2},\ldots,a_{\tau_l})}}\ \sum_{\substack{(\hat{a}_{i_1},\hat{a}_{i_2},\ldots,\hat{a}_{i_s})\in A\\ i_j\neq\tau_r}}\ \prod_{\substack{j=1\\ i_j\neq\tau_r}}^{s} P(\hat{a}_{i_j}|a_{i_j}). \tag{4.26}$$

We note that all the tuples $(\hat{a}_{i_1}, \hat{a}_{i_2}, \ldots, \hat{a}_{i_s}) \in A$ with $i_j \neq \tau_r$, which are used in the summation above, are different for different tuples $(\hat{a}_{\tau_1}, \hat{a}_{\tau_2}, \ldots, \hat{a}_{\tau_l})$. To prove this, suppose the opposite; that is two different tuples $(\hat{a}_{\tau_1}, \hat{a}_{\tau_2}, \ldots, \hat{a}_{\tau_l})$ and $(\hat{a}_{\sigma_1}, \hat{a}_{\sigma_2}, \ldots, \hat{a}_{\sigma_l})$ correspond to the same $(\hat{a}_{i_1}, \hat{a}_{i_2}, \ldots, \hat{a}_{i_s})$. If we erase the positions $(\hat{a}_{\tau_1}, \hat{a}_{\tau_2}, \ldots, \hat{a}_{\tau_l})$ it will be impossible to recover the code word in the RS code, this contradicts the fact that a RS code can correct $n_2 - k_2$ erasures, whereas above only $n_2 - k_2 - l$ positions are erased. Therefore

$$\sum_{\substack{(\hat{a}_{\tau_1},\hat{a}_{\tau_2},\ldots,\hat{a}_{\tau_l})\\ \neq(a_{\tau_1},a_{\tau_2},\ldots,a_{\tau_l})}}\ \sum_{\substack{(\hat{a}_{i_1},\hat{a}_{i_2},\ldots,\hat{a}_{i_s})\in A\\ i_j\neq\tau_r}}\ \prod_{\substack{j=1\\ i_j\neq\tau_r}}^{s} P(\hat{a}_{i_j}|a_{i_j}) \leq 1. \tag{4.27}$$

It is easy to show that

$$\max_{p^{(\tau_r)}}\left\{\left(p^{(\tau_r)}\right)^{d_1-t}\left(1-p^{(\tau_r)}\right)^{n_1-d_1+t}\right\} \leq \left(\frac{d_1-t}{n_1}\right)^{d_1-t}\left(1-\frac{d_1-t}{n_1}\right)^{n_1-d_1+t}. \tag{4.28}$$

Substituting (4.27) and (4.28) into (4.26) and taking into account that the outer sum contains at most $\binom{n_2}{l}$ terms, the theorem follows.

Another upper bound for an LBC with non-random parameters $p^{(1)}, p^{(2)}, \ldots, p^{(s)}$ can be obtained if we use a bound of guaranteed multiplicity for the error detecting by outer code after correction of at most $n_2 - k_2 - l$ erasures

$$P_{\text{ue}}(C, LBC\text{-}s) \leq \sum_{i=l+1}^{n_2}\binom{n_2}{i} P_{\text{wc}}^{(t)}(C_1, BSC)^i \left(1 - P_{\text{wc}}^{(t)}(C_1, BSC)\right)^{n_2-i}, \tag{4.29}$$

where C_1 is the inner code.

The bound (4.29) is sometimes superior, sometimes inferior to the bound in Theorem 4.2.6.

Example. Let C be a concatenated code with parameters $n_1 = 31$, $k_1 = 16$, $d_1 = 7$, $t = 0$, $n_2 = 10$, $k_2 = 5$, $l = 1$. Then Theorem 4.2.6 gives $P_{\mathrm{ue}} \leq 6.4 \cdot 10^{-7}$ and (4.29) gives $P_{\mathrm{ue}} \leq 4.2 \cdot 10^{-8}$. However, if we change l to 2 and keep the other parameters we get $P_{\mathrm{ue}} \leq 1.8 \cdot 10^{-13}$ from Theorem 4.2.6 and $P_{\mathrm{ue}} \leq 3.4 \cdot 10^{-12}$ from (4.29).

Remark. A main advantage of the bound in Theorem 4.2.6 is that it depends only on the parameters n_1, k_1, d_1, n_2, k_2, t, and l. The bound is valid for any LBC-s where s is a multiple of n_1. To calculate (4.29) we need to calculate or estimate $P_{\mathrm{wc}}^{(t)}(C_1, BSC)$. Further, the bound is valid only for LBC with non-random parameters.

Error Detection After Channel Transform

In Chapter 1 we considered different deterministic and stochastic transforms of a channel. After outer block interleaving, we can use, for the RFC model, the bound obtained from (4.17) when r is changed to r^s. Similarly, for the FMC model we can use (4.18) and (4.19) where $\mathbf{A}_0$ and $\mathbf{A}_1$ are changed to $\mathbf{A}_0(\mathbf{A}_0 + \mathbf{A}_1)^{s-1}$ and $\mathbf{A}_1(\mathbf{A}_0 + \mathbf{A}_1)^{s-1}$ respectively. Starting from any binary symmetric channel with memory and making the interleaving interval sufficiently large, we obtain a BSC with symbol error probability equal to the average symbol probability of the original channel. We can therefore use all the theory developed in Chapter 3.

After stochastic outer block interleaving, we have a quasi binomial channel and therefore

$$P_{\mathrm{ue}}(C) = \sum_{m=1}^{n} A_m(C)\tilde{p}(n, m), \tag{4.30}$$

where $\tilde{p}(n, m)$ is given by (1.38).

The disadvantage of outer block interleaving is, as a rule, a decrease in $P_{\mathrm{ef}}(n)$ and consequently in the throughput efficiency of the feedback system. (See further in Chapter 5). Using inter block pseudorandom interleaving keeps the value of $P_{\mathrm{ef}}(n)$ unchanged. After such a transformation we also have a quasi binomial channel with the same multiplicity function $P(n,m)$ as the original channel. In this case we have $\tilde{p}(n,m) = P(n,m)/\binom{n}{m}$ in (4.30).

For practical application it is most interesting to combine the two transforms SATC and SMTC, because this allows us to design very simple error detecting codes with guaranteed upper bounds on the probability of undetected error. Using (1.44) we get

$$P_{\text{ue}}(C, SATC + SMTC) = \frac{2^k - 1}{2^n - 1}(1 - P_{\text{ef}}(n)) \tag{4.31}$$

for any binary systematic $[n,k]$ code on *any binary channel* after transform. This relation coincides with (4.3) for $q = 2$.

Remark. It would be possible to define q-ary SATC and SMTC and use it with q-ary systematic codes for error detection. However, such a generalization has no clear advantage over the binary case in terms of $P_{\text{ue}}(C)$ and it would be more complex to implement.

The simplest code which can be used with SATC and SMTC is a trivial code with all zero check symbols. If we receive a block with some non-zero check symbol we detect an error. This simple example gives P_{ue} given by (4.31). It is very important that $P_{\text{ef}}(n)$ remains the same as in the original channel and therefore the throughput efficiency is unchanged.

We mentioned in Chapter 1 that an actual implementation of SATC and SMTC are pseudo-random transforms and that this requires synchronization between the sending and the receiving ends. This can be avoided using a pure random generator at the transmitting end and transmitting the random elements together with the information. However, this results in a halving of the rate and a decrease in $\tilde{P}_{\text{ef}}(n)$ after the transform to $P_{\text{ef}}(2n)$ in the original channel. It will depend on the requirements for the information system whether the use of a random or a pseudo-random generator is best. One added advantage offered by the pseudo-random generator is the possibility to protect the information from undetected deception. This can be done using a good generator similar to a stream cipher generator and secret keys.

After SATC and SMTC we have a uniform probability distribution of the error patterns. Therefore, it is not possible to use combined error correction and detection. However, we can use two codes, an $[n,k]$ code C for error detection and an $[n',n]$ code C' for error correction, and place the SATC and SMTC between the encoder for C and the encoder for C', and similarly at the receiving end. Thus C' is used on the original channel and C on the transformed channel. This approach is inferior to using a single code for error correction and detection; however it avoids the problem of finding a t-good code. The problem of error detecting code optimization will be considered in Chapter 6.

As was remarked in Chapter 1, using a linear shift register of length n with a primitive feedback polynomial will secure that (1.44) and (4.31) are satisfied. For $Q(N)$, the probability of receiving N blocks without undetected errors, we are not guaranteed that

$$Q(N) = 1 - [P_{\mathrm{ue}}(C, SATC + SMTC)]^N. \tag{4.32}$$

However, we have the following lower bound:

$$Q(N) \geq 1 - N \cdot P_{\mathrm{ue}}(C, SATC + SMTC). \tag{4.33}$$

Moreover, it can be shown that there exists a channel which gives equality in (4.33). The relation (4.33) is suitable for estimating the reliability for message sections if $P_{\mathrm{ue}}(C, SATC + SMTC)$ is small.

Remark. Another way to do coding is by the use of hashing. A hash function in our context is a transformation $h : GF(2)^k \to GF(2)^r$. An information block $\bar{x} \in GF(2)^k$ is encoded into $(\bar{x}|\bar{z})$ where $\bar{z} = h(\bar{x})$. In the special case when h is a linear transformation, we get a systematic linear code as defined before. If $(\bar{x}|\bar{z})$ is sent and $(\bar{x}'|\bar{z}')$ is received, we get an undetected error if $\bar{x}' \neq \bar{x}$ and $h(\bar{x}') = \bar{z}'$. We consider the average probability of undetected error when h is choosen from some set H randomly and with uniform distribution.

The first case we consider is the situation when $\bar{z}$ is assumed to be error free (this can model the situation with a protocol with information feedback on a noiseless return channel or a situation where $\bar{z}$ is stored in a perfect memory). If the channel is described by the transition probabilities $P(\bar{x}'|\bar{x})$, where $\bar{x}, \bar{x}' \in GF(2)^k$, we get

$$E\Big(P_{\mathrm{ue}}(\bar{x})\Big) = \sum_{\bar{x}' \neq \bar{x}} P(\bar{x}'|\bar{x}) P\{h(\bar{x}) = h(\bar{x}')\}. \tag{4.34}$$

A set H of hash functions $GF(2)^k \to GF(2)^r$ is called *universal* if for all $\bar{x} \neq \bar{x}'$ we have

$$P\{h(\bar{x}) = h(\bar{x}')\} = \frac{\#\{h \in H \mid h(\bar{x}) = h(\bar{x}')\}}{\#H} \leq \frac{1}{2^r}.$$

For a universal set of hash functions we get, from (4.34), that

$$E\Big(P_{\text{ue}}(\bar{x})\Big) \le \frac{1}{2^r}\Big(1 - P_{\text{ef}}(k)\Big), \tag{4.35}$$

where $P_{\text{ef}}(k)$ is the probability of an error free information block.

An example of a universal set of hash functions is the following: we identify $GF(2)^k$ and $GF(2^k)$ in the natural way. We let $H = GF(2^k)$. For $\bar{h}, \bar{x} \in GF(2^k)$, let $\bar{h} \cdot \bar{x} = (u_1, u_2, \ldots, u_k)$ (where $\cdot$ denotes multiplication in $GF(2^k)$), and define

$$\bar{h}(\bar{x}) = (u_1, u_2, \ldots, u_r). \tag{4.36}$$

The second case we consider is a channel with corruption, with additiv noise, of both $\bar{x}$ and $\bar{z}$. Then we obtain the following relation for the average probability of undetected error:

$$E\Big(P_{\text{ue}}((\bar{x}|\bar{z}))\Big) = \sum_{(\bar{e}_{\bar{x}}|\bar{e}_{\bar{z}}) \ne \bar{0}} P(\bar{e}_{\bar{x}}|\bar{e}_{\bar{z}}) P\{h(\bar{x} + \bar{e}_{\bar{z}}) = h(\bar{z} + \bar{e}_{\bar{x}})\}. \tag{4.37}$$

In particular, for the set of hash functions defined by (4.36) it is easy to show that

$$P\Big\{\bar{h}(\bar{x} + \bar{e}_{\bar{z}}) = \bar{h}(\bar{z} + \bar{e}_{\bar{x}})\Big\} = P\Big\{\bar{h} \cdot \bar{e}_{\bar{x}} = \bar{e}_{\bar{z}}\Big\} = \frac{1}{2^r}$$

for all $\bar{e}_{\bar{x}} \ne \bar{0}$. Substituting this into (4.37) we get

$$E\Big(P_{\text{ue}}(\bar{x})\Big) = \frac{1}{2^r}\Big(1 - P_{\text{ef}}(k)\Big). \tag{4.38}$$

We see that both (4.35) and (4.38) are very close to (4.31). Hashing is also very close to the stochastic multiplication transform. An actual implementation of hashing can be done in the form of identical pseudo-hash functions at the sending and receiving ends.

4.3 CONCLUSIONS

In general it is hard to calculate $P_{\text{ue}}(C)$ for a linear code on the BSC as we have seen in Chapter 3. It is not surprising that it is usually harder to calculate for channels with memory. Corollary 4.2.1 shows that for any channel with memory, having additive noise, there exists an $[n, k; q]$ code with probability of undetected error less then $q^{-(n-k)}$. On the other hand, Corollary 4.2.3 shows that in general there will not exist codes with $P_{wc}(C, K) \leq (q^k - 1)/q^n$. We can overcome this problem by changing from deterministic to stochastic coding (that is, using channel transforms). The latter approach is especially good since it is not difficult to find good codes. A code satisfying (4.31) can be obtained simply by having all the check symbols equal to zero. But we should then create a special synchronization system to provide the same transforms at the transmitting and at the receiving side. We can avoid this requirement using a pure random generator, but at the cost of a decrease in the data rate by a factor 2. Combined correction and detection is also possible with stochastic transforms, but again at a reduced rate compared with using a single optimal code for both correction and detection.

We can use stochastic transforms not only on channels with memory, but also on BSC. This may be of interest if we can not calculate $P_{ue}(C, BSC)$ or find a good code with given parameters. We return to the problem of searching for practical optimal codes in Chapter 6.

For a number of channel models we can find the probabilities of undetected error for coding without stochastic transforms. These are the CC and REF, see formulas (4.16), (4.17), (4.23). The bound (4.13) is valid for any channel with additive noise, but as a rule it is weaker. As for (4.12), it is a better bound, but it requires knowledge of the $p(n, i)$ function and this is not known in most cases.

Theorem 4.2.6 can be very useful if we are dealing with VPC satisfying the additional condition to be strongly correlated for the time required to transmit a code block of length n_1. Then we can construct a concatenated code C with inner code of block length n_1 such that $P_{ue}(C, VPC)$ satisfies Theorem 4.2.6 and (4.29).

We will continue our discussion about searching for practical optimal codes for channels with memory in Chapter 6, because a decision about choice of a code must take into account not only P_{ue}, but also the throughput efficiency provided by this code and the protocol for the feedback system.

4.4 COMMENTS AND REFERENCES

4.1. Freiman [63] proved that the largest perfect C of length n is obtained by taking all vectors of weight $\lfloor \frac{n}{2} \rfloor$.

The systematic perfect code given is due to Berger [36]. Freiman [63] proved that no perfect systematic code has less redundancy.

Theorem 4.1.1 is new. Freiman suggested using the code with $\{a_1, a_2, \ldots a_k\}$ $\{\lfloor n/2 \rfloor \pm i(a+1) \mid 0 \leq i \leq n/(2*(a+1))\}$ to detect up to a asymmetric errors. Note that the example given in Section 4.1 shows that this construction may not be optimal. Freiman also have a suggestion for a systematic code to detect up to a asymmetric errors.

A channel related to the Z-channel is the *unidirectional channel.* For this channel, if there are errors in a code word, they are all $0 \to 1$ errors or all $1 \to 0$ errors. A number of codes have been constructed for correction or detecting of such error-patterns. Also there are many constructions of codes correcting t symmetric errors (where t usually is 1) and detecting all unidirectional errors. For more information on these codes, we refere the reader to the book by Rao and Fujiwara [20].

4.2. Most of the quoted results are given in Korzhik and Fink [10].

The results on the worst-case probability of undetected error on the LBC are taken from Kløve [110].

Theorem 4.2.6 is due to Korzhik and Dzubanov [117].

4.3. The idea to combine a simple error detecting code with stochastic transforms is due to Korzhik, Osmolovskii, and Fink [120]; see further Korzhik and Fink [10] and Fink and Mukhametshina [61].

A similar approach to avoid deception of false information has been proposed by Tanaka and Kaneki [187].

Kuznetsov et al. [127] considered the probability of undetected error of linear block codes on the channel with memory described by Markov distributions. Their model is the particular case of the FPMC model described in Chapter 1 obtained by taking diagonal matrices $\mathbf{A}_0$ and $\mathbf{A}_1$. For this model they obtained interesting bounds on the average probability of undetected error, similar to the results for the BSC given in Section 3.2. They also give a recursive technique for computing the probability of undetected error based on an exhaustive search of the dual code.

A recursive technique for computing P_{ue} for shortend cyclic codes on the LBC-1 channels is given by Agarwal and Ivanov [24]. The complexity of the computation is $O\big(n|g|2^{r+|g|}\big)$, where n is the block length, r the redundancy, and $|g|$ is the number of non-zero coefficients of a generator polynomial g. In particular, this approach can be used to calculate $P_{\text{ue}}(C, BSC_p)$ for the code C generated by g without an explicit computation of the weight distribution of the code.

Universal hash functions were introduced by Carter and Wegman [44], [199] (they called them universal$_2$ hash functions).

5

PROTOCOLS FOR INFORMATION TRANSMISSION SYSTEMS WITH FEEDBACK

5.1 INTRODUCTION.

In the previous chapters we considered the use of linear codes for error detection (may be with partial correction) and the error detecting capability of these codes. The results apply to the so called simplex communication systems, where we have only a one way channel from user A to user B. When an error is detected in a block, it is marked as unreliable. The further processing may be vary — for example, the block may be erased or used critically and so on. A scheme where we do error correction jointly with error detection is known as a *forward-error-control* (FEC) scheme.

But it is more natural to use error detecting codes in the so called *feedback systems* (FS). In this case we have a duplex channel — from A to B and from B to A. It is very important that the return channel can be used to improve the information transmission over the forward channel. (Otherwise we will only have two simplex channels). It is usual to call these schemes *automatic-repeat-request* (ARQ) schemes, because if errors are detected in the received vector, the transmitter is asked to retransmit the same code vector, and this retransmission is repeated until the code vector is successfully received. Actually there are many different procedures for exchange of information between A and B. We call such a procedure a *protocol of data exchange* or briefly a protocol.

The forward and return channels can be described, in general, by different channel models (see Chapter 1), and the errors in these channels can be interdependent. However, in this book we consider the situation where both channels have the same model and are mutually independent. In the extreme case, when the return channel is used to improve the forward channel only, and it is modeled by a noiseless channel, we have a *perfect feedback system.*

The key point of all protocols is taking into account the *round-trip delay* of a given two-way channel. This is defined as the sum of the time delays in the forward and return channels. In the extreme case, when the time of the signal propagation in both channels are close to zero, we have an *instantaneous feedback system.*

Remark. Even for an instantaneous feedback system we can not receive an acknowledgement or a negative acknowledgement immediately after a code block transmission, because it is necessary to receive at least one symbol over the return channel.

This chapter starts with a description of general efficiency criteria, which are typical for FS, when the time of the final block reception by a user is a random value. We concentrate on the main protocols, which promise to have good efficiency.

A typical presentation of a protocol is as follows: Description of the algorithm, estimation of its efficiency, first for pure detection, and then for the so called hybrid algorithms (joint detection and correction). In general we will use BSC as the main model of both the forward and return channels but in all cases we will consider some channel models with memory.

There are two main types of FS. The first is the so called *decision feedback systems,* which makes decisions about retransmission at the receiving end. The other type of the FS is the *information feedback systems,* which makes similar decisions at the transmitting end. We will discuss, as a rule, the first type, and only at the end of this chapter touch the second type.

Remark. We will not describe all the more than twenty proposed protocols for the FS. Only the main protocols have been chosen, those which are used in practice now or which in our view have a good potential for engineering designs in the future.

5.2 THE MAIN CRITERIA OF FS EFFICIENCY

It is usual to distinguish between two groups of measures for FS efficiency:

- reliability of information transmission;
- delivery time of messages.

Together, they give a *time-reliability* (TR) description of information transmission systems. It is necessary to take into account the complexity of determining these TR. But we will take this notion only as some bounding condition. If we remove this requirement completely it will result in rejection of any FS, at least for channels without memory, because it has been proved by Dobrushin [55] that the capacity of a channel with FS and without memory is equal to the capacity of the same channel without feedback.

Remark. As we pointed out in Chapter 4, if we have a CC channel model, then the use of the FS is useful in spite of its complexity, because we can guarantee any given reliability for any channel states. This is impossible for forward error control only.

Let us consider at first the reliability criterion. We have derived many relations to calculate or to estimate the probability of undetected errors (see Chapters 3 and 4). The probability of undetected error seems to be suitable as a measure of reliability. However, there are three additional circumstances which have to be taken into account:

(i) Different codes have different numbers of information symbols in the code blocks.

(ii) The probability of undetected errors determines only the error rate at the output of the channel, but does not give directly the error rate of the information transmitted to a user.

(iii) Besides undetected (substitution) errors in the FS, we also get other types of errors, namely, *insertions* and *deletions.*

It is possible to consider these types of errors as ordinary undetected errors, but in our opinion it is better to distinguish between them, because some information systems can have special requirements for this kind of errors. (This separation into three kinds of errors also simplifies the calculations).

We define the following limits in probability. We assume that the limits exist, this is always the case for the BSC.

The probability $P_{\text{ef}}(n)$ of an *error free block* of length n is the value to which the ratio of the number of correctly received blocks N_{c} to the number of blocks N transmitted over the channel, is converging in probability, that is, for any $\epsilon > 0$ we have

$$\lim_{N\to\infty} P\left\{\left|\frac{N_{\text{c}}}{N} - P_{\text{ef}}(n)\right| > \epsilon\right\} = 0. \tag{5.1}$$

The probability P_{d} of *detected errors* is the value to which the ratio of the number of blocks N_{d} with detected errors to the number of blocks N transmitted over the channel is converging in probability

$$\lim_{N\to\infty} P\left\{\left|\frac{N_{\text{d}}}{N} - P_{\text{d}}\right| > \epsilon\right\} = 0 \text{ for any } \epsilon > 0. \tag{5.2}$$

The probability P_{ue} of the *undetected errors* is the value to which the ratio of the number of blocks with undetected errors at the output of the channel N_{ue} to the number of blocks N transmitted over the channel is converging in probability

$$\lim_{N\to\infty} P\left\{\left|\frac{N_{\text{ue}}}{N} - P_{\text{ue}}\right| > \epsilon\right\} = 0 \text{ for any } \epsilon > 0. \tag{5.3}$$

The *final error probability* P_{f} is the value to which the ratio of the number $\hat{N}_{\text{ue}}$ of received blocks with undetected errors to the total number of received blocks $\hat{N}$ is converging in probability

$$\lim_{N\to\infty} P\left\{\left|\frac{\hat{N}_{\text{ue}}}{\hat{N}} - P_{\text{f}}\right| > \epsilon\right\} = 0 \text{ for any } \epsilon > 0. \tag{5.4}$$

The *probabilities of insertion* (P_{in}) or *deletion* (P_{de}) are the values to which the ratios of the numbers $\hat{N}_{\text{in}}$ or $\hat{N}_{\text{de}}$ of blocks, which are inserted or deleted respectively, to the number of received blocks $\hat{N}$ are converging in probability

$$\lim_{N\to\infty} P\left\{\left|\frac{\hat{N}_{\text{in(de)}}}{\hat{N}} - P_{\text{in(de)}}\right| > \epsilon\right\} = 0 \text{ for any } \epsilon > 0. \tag{5.5}$$

The probability of *correct transmission of an information block* to a user, P_c, is the value to which the ratio of the number of information block-sessions N_c received by the user without any errors to the number of information block-sessions N_0 transmitted over the channel, is converging in probability

$$\lim_{N\to\infty} P\left\{\left|\frac{N_c}{N_0} - P_c\right| > \epsilon\right\} = 0 \text{ for any } \epsilon > 0. \tag{5.6}$$

Note that P_c is the most important value if we want to transmit an information session without any errors, but possibly with insertions. Therefore, we also define the probability of information shift P_{sh} as the value

$$P_{sh} = P_{in} + P_{de}.$$

It follows from the definitions that for any FS protocol

$$P_f = \frac{P_{ue}}{P_{ef} + P_{ue}}. \tag{5.7}$$

There are two typical groups of users of information systems. The first group requires absence of any errors during a communication session consisting of N_0 information blocks. It does not matter how many errors we have at the session if there are any at all. This situation is the usual for data communication.

The second group requires minimization of the number of errors during a transmission session. That situation is typical for digitized speech transmission and text transmission.

It is natural to define reliability for the first group of users as the probability $P_e(N_0)$ that a whole session has at least one error. For the BSC we get

$$P_e(N_0) = 1 - P_c^{N_0}. \tag{5.8}$$

For a channel with memory we can give only the upper bound

$$P_e(N_0) \leq N_0\,(1 - P_c)\,. \tag{5.9}$$

It is also necessary to give conditions on the probability $P_{sh}(N_0)$ of information shifts for a session.

Let us consider the BSC without any coding and feedback and take P_{eq} as the symbol error probability of such a channel. Given $P_e(N_0)$ we would like to find the P_{eq} (the equivalent symbol probability), which gives with the ordinary BSC the same value of $P_e(N_0)$ as the given error detecting and feedback system. This results in the equation

$$1 - (1 - P_{eq})^{N_0 k} = 1 - P_c^{N_0} \tag{5.10}$$

Solving (5.10) with respect to P_{eq} we obtain

$$P_{eq} = 1 - P_c^{1/k}$$

If P_f is small enough we get from (5.10)

$$P_{eq} \approx (1 - P_c)/k. \tag{5.11}$$

(We will very often use P_{eq} given by (5.10) or (5.11) as a criterion of FS reliability for information transmission over the BSC).

For the second group of users it is natural to define reliability as the averaged symbol error probability P_{av} given the error detecting code. Its determination in the general case results in complex calculations involving the Hamming weight distributions, but we will use the approximation for this value given by Clark and Cain [5, page 23]

$$P_{av} \approx \frac{d}{n} P_f \tag{5.12}$$

where d is the minimum code distance for the given code used for error detection.

Let us now consider the delivery time criteria. Since the time of the message portion reception is a random value, the most general description of it will be by a *distribution function.* For block codes this distribution function $F_k(S)$ is defined as the probability to deliver k information symbols to the user after transmission of at most S code blocks over the channel. (If the channel has no memory, then it is easy to find the probability to deliver any portion of information bits given $F_k(S)$).

There are two simpler characteristics of the delivery time, the expected number of channel symbols required to deliver one information symbol to a user, and its variance. The first, and most important of these, is called the ***throughput efficiency*** (TE). In other words, it is defined as the ratio of the average number of information symbols successfully received by the user to the total number of symbols transmitted over the channel. This results in the following definition of the throughput efficiency η via convergence in probability

$$\lim_{N\to\infty} P\left\{\left|\frac{k\hat{N}}{\tilde{n}N} - \eta\right| > \epsilon\right\} = 0 \text{ for any } \epsilon > 0 \tag{5.13}$$

where

- N is the number of code blocks transmitted over the channel,
- $\hat{N}$ is the number of code blocks accepted by the user,
- $\tilde{n}$ is the number of channel symbols transmitted over the channel during the time between the beginnings of two adjacent code blocks,
- k is the number of information symbols in the code blocks.

With the knowledge of η we can find the average time $T_{\text{av}}(N_0)$ to transmit N_0 information digits on a given FS, namely

$$T_{\text{av}}(N_0) = N_0/v_c \cdot \eta \text{ (seconds)} \tag{5.14}$$

where v_c is the channel rate (the number of channel symbols transmitted over the channel per second).

The throughput efficiency is usually a good measure describing the delivery time characteristics, the exception may be when the transmitted data has to delivered within a deadline or otherwise rapidly looses its value over time.

It is often more convenient to extract the code rate $R_c = k/\tilde{n}$ from η. We have

$$\eta = R_c \cdot \eta_0 \tag{5.15}$$

where η_0 is determined by

$$\lim_{N\to\infty} P\left\{\left|\frac{\hat{N}}{N} - \eta_0\right| > \epsilon\right\} = 0 \text{ for any } \epsilon > 0. \tag{5.16}$$

To find $P_{\rm f}, P_{\rm in}, P_{\rm de}$, and η it is necessary to know the protocol used, the code used for error detection and correction, and the channel model. We will consider these problems in the next two sections for decision feedback systems and information feedback systems, respectively.

5.3 BASIC PROTOCOLS FOR DECISION FEEDBACK SYSTEMS

There are a number of different protocols for FS and more are introduced. Some of them are chosen as international standards, for example X.25. To describe them all is beyond the scope of this book. We consider here three basic types of protocols: the ***stop-and-wait*** ARQ, the ***go-back-M*** ARQ, the ***selective repeat*** (or ***address-request***) ARQ as well as some other modern ***iterative*** protocols. It is very important when we design a protocols to understand whether we know exactly the round-trip delay in the channel or we know only an upper bound for this value. We have two possibilities for the return channel; it can be used to improve the information transmission over the forward channel only or for the transmission of its own information as well.

The exposition of the different protocols will be as follows:

- description of the algorithm;
- estimates of TE for error detection on the BSC;
- estimates of TE for joint error detection and correction (hybrid methods) on the BSC;
- estimates of TE for some channels with memory.

The Stop-and-Wait ARQ Protocol

This is the simplest protocol, and it is the first that comes to mind. The transmitter sends a code block to the receiver and waits for an acknowledgement (ACK) or negative acknowledgement (NAK) signal from the receiver. If ACK is received, then the transmitter sends the next code word and if NAK is received, then the transmitter resends the same code word. Retransmissions continue until ACK is received by the transmitter.

The main defect of this protocol is the idle time spent waiting for an acknowledgement for each transmitted code word. Denote this idle time by t_{it}. It is given by the following relation

$$t_{\text{it}} = t_{\text{rt}} + t_{\text{ac}} \tag{5.17}$$

where

- t_{rt} is the round trip delay,
- t_{ac} is the time of ACK (NAK) block transmission. (We neglect the processing time).

This protocol can be used for duplex communication. Two blocks are transmitted in each direction (one with information and the other with ACK or NAK signals about information received over the opposite channel) or one block contains both information symbols which are intended for the forward channel, and service symbols (ACK or NAK) for the return channel.

To provide protection against insertion this protocol can be modified in the following manner. If some information block is received without detected errors, it is compared with the previous information block, and when they coincide the last received block is erased (without transmission to the user) and over the return channel ACK is transmitted. When they do not coincide the accepted block is transmitted to the user, and over the return channel ACK is again transmitted. Before transmission of the next information block, it is compared with the previous one, and if they coincide, we transmit instead of the next block a special information block, which signifies a repetition of the previous block.

In Figure 5.1, the state operation diagram corresponding to this protocol with separated information and service blocks is given.

From (5.4), (5.6), and the description of this protocol we have (see also Benice and Frey [35])

$$P_{\text{c}} = \frac{P_{\text{ef}}}{P_{\text{ef}} + P_{\text{d}}\ P'_{\text{ue}} + P_{\text{ue}} \cdot P'_{\text{ef}}} \tag{5.18}$$

where " ′ " means the corresponding probability in the return channel. This expression has the following lower bound for similar forward and return channels

$$P_{\text{c}} \geq 1 - P_{\text{f}}, \tag{5.19}$$

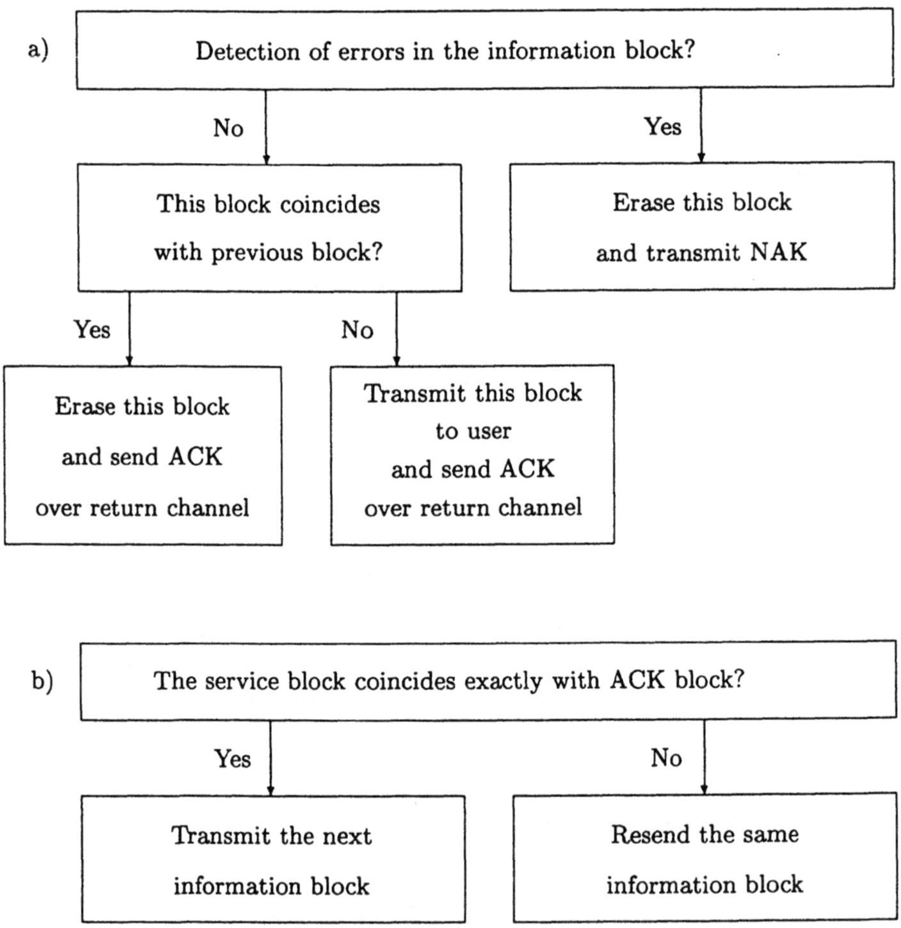

Figure 5.1 The block diagram of the state operation for the stop-and-wait ARQ protocol for a) the information block & b) the service block.

and we will have equality in (5.19) for a noiseless return channel. (In general the difference $P_c - (1 - P_f)$ is P_{ue}^2).

Remark. The relation (5.7) is true for any channels with memory but for CC (see Chapter 1) we can give a simpler bound using its definition and inequality (1.8)

$$P_f \leq \frac{P_{wc}(C, 0, 1/2)}{P_{wc}(C, 0, 1/2) + (1 - \bar{p})^n} \tag{5.20}$$

where

- $P_{wc}(C, 0, 1/2)$ is the worst-case probability of undetected error for a given code C on the BSC,
- $\bar{p}$ is the average symbol error probability in CC (see Chapter 1).

If we do not know the average symbol error probability $\bar{p}$, we are forced to consider the worst case with $\bar{p} = 1/2$ which gives the trivial result $P_f \leq 1$.

To provide a guaranteed reliability in any case, we should limit the number of retransmissions by some moderate number α. If more than this number of retransmissions are asked for, we conclude that the channel is too bad and we should refrain from using it or somehow "correct" it. We will then have the following upper bound on P_f:

$$P_f \leq \alpha P_{wc}(C, 0, 1/2).$$

For the BSC with symbol error probability p and for pure error detection we get from (5.7)

$$P_f = \frac{P_{ue}}{P_{ue} + (1 - p)^n} \tag{5.21}$$

where n is the block length, and then P_{eq} and P_{av} can be found by substituting (5.19) and (5.21) into (5.11) and (5.12) respectively.

For separately transmitted information and service blocks it is natural to take a 0-block of length n' as an ACK word and a 1-block of length n' as a NAK word (or vice versa). According to Fig. 5.1 we use such an algorithm when a service block is recognized as ACK if and only if it contains n' zeros at the receiving end, otherwise it is recognized as a NAK block. Therefore a block will be deleted either when ACK is received in place of NAK or we have undetected errors and the block coincides with the previous information block. The probability of the first outcome is equal to the probability of errors in all n' symbols of the service block and of the second outcome is at most the probability of undetected errors. Then taking (5.1) and (5.5) we get for the BSC

$$P_{\text{de}} \leq \frac{p^{-n}}{(1-p)^n + P_{\text{ue}}} + \frac{P_{\text{ue}}}{(1-p)^n + P_{\text{ue}}}. \tag{5.22}$$

If we choose n' satisfying $2^{-n'} < P_{\text{ue}}$ for any $0 \leq p \leq 1/2$, then we obtain, from (5.22),

$$P_{\text{de}} \leq 2P_{\text{f}}. \tag{5.23}$$

To find the probability of an insertion (P_{in}), we note that a repeated block will be given to the user when the following two conditions are true: the ACN block has been transformed to the NAK block and either the confirmed or the resent block have undetected errors. The probability of such an event for the BSC is $(1-(1-P_{\text{ue}})^2)(1-(1-p)^{n'})$.

From (5.5) we then get

$$P_{\text{in}} \leq \frac{(1-(1-P_{\text{ue}})^2)\left(1-(1-p)^{n'}\right)}{(1-p)^n + P_{\text{ue}}} \leq 2P_{\text{f}}. \tag{5.24}$$

The relations (5.23) and (5.24) show that for this protocol the probability of an information shift $P_{\text{sh}} \leq 4P_{\text{f}}$ and it has the same order of magnitude as P_{f}. If we want to have P_{sh} much less than P_{f} it is necessary to use other protocols.

It was mentioned earlier that this protocol has a modification with joint transmission of information and service symbols. If we use a systematic binary $[n, k_1]$ code, $k = k_1 - 1$ symbols are information symbols and one symbol is a service symbol (0 corresponds to ACK and 1 corresponds to NAK). The algorithm for this modification is practically the same as the original. There is only one difference, every code block with a detected error is considered as a block containing a NAK bit and this results in retransmission of the block in the opposite direction.

It is easy to see that formulas (5.19), (5.23), (5.24) remain true in this case.

Now we consider delivery time characteristics for this protocol with joint transmission of information and service symbols. Denote by Δ the probability of the event that the given transmission of the block will be the last attempt to convey it. This may happen in three cases: either this block is error free and it will not be resent, it is accepted with undetected errors and will not be resent, or it is received with detected errors but will not be repeated, because NAK is transformed into ACK due to an error in the return channel.

These events are disjoint and therefore Δ is equal to the sum of their probabilities

$$\Delta = P_{\text{ef}} P'_{\text{ef}} + P_{\text{ue}} P'_{\text{ef}} + P_{\text{d}} P'_{\text{ue}}. \tag{5.25}$$

where " ′ " denotes the corresponding probabilities in the return channel.

The distribution function $F_k(S)$ is given by the following relation for the BSC

$$F_k(S) = \Delta \sum_{i=0}^{S-1} (1-\Delta)^i = 1 - (1-\Delta)^S. \tag{5.26}$$

The average number of code blocks necessary to be transmitted over a BSC channel to get one information block accepted is the expectation of S

$$\bar{S} = \sum_{i=1}^{\infty} i\Delta(1-\Delta)^{i-1} = 1/\Delta. \tag{5.27}$$

Because the values P_{ue} and P'_{ue} usually are small compared to P_{ef} and P'_{ef} we can use the following rather precise lower bound for Δ on the BSC

$$\Delta \geq P_{\text{ef}} P'_{\text{ef}} = [(1-p)(1-p')]^n. \tag{5.28}$$

Then the lower bound of the TE for this protocol has the following expression

$$\eta \geq \frac{(k-1)[(1-p)(1-p')]^n}{2n + \lceil t_{\text{rt}}/\tau \rceil} \tag{5.29}$$

where τ is the symbol interval in the channel.

We can see from (5.29) that even for very small values of t_{rt}, for which $\lceil t_{\text{rt}}/n \rceil \ll n$ in the denominator of (5.29), we obtain an additional term equal to n because of the idle time. This value can be lowered if we use a shorter block of length $n' < n$ for transmission of service symbols but this does not solve the problem. This protocol is very inefficient in terms of TE, especially for large values of t_{rt}. Therefore, we will not consider channel models with memory for this protocol. As for hybrid schemes (joint error correction and detection) for the BSC we only remark that all formulas remain the same, if we everywhere change $P_{\text{ef}} = (1-p)^n$ to

$$P_{\text{ef}}^{(t)} = \sum_{i=0}^{t} \binom{n}{i} p^i (1-p)^{n-i}, \tag{5.30}$$

where t is error correcting capability of the code used (see Chapters 2 and 3).

The Go-Back-M ARQ Protocol

To remove the main defect of the previous protocol, namely the idle time, the information blocks could be transmitted continuously and the acknowledgment or negative acknowledgment are not attributed to the previous block but to some earlier transmitted block, whose position can be found given the round-trip delay. In this case all relations derived for the stop-and-wait protocol remains true after letting $n_{\mathrm{it}} = 0$.

But such an approach has two significant drawbacks: first, it requires a memory with random access, and second, the round-trip delay in the channel to be used for information transmission must be known exactly. We will return to this approach in the next subsection, but first we consider a simpler method, which requires only a stack memory and knowledge of an upper bound of the round-trip delay. Choose an integer M satisfying the following inequality

$$M \geq \lceil (n_{\mathrm{rt}} + n_{\mathrm{an}} + n_{\mathrm{s}})/n \rceil \tag{5.31}$$

where

- n_{s} is the number of symbols needed to transmit the ACK or NAK blocks,
- n_{an} is the number of symbols received during the period needed for the analysis of a block.

When receiver B detects errors he makes three actions:

- transmits an NAK block over the return channel to A,
- rejects the block with detected errors,
- closes the receiver for $M - 1$ blocks after the block with detected errors regardless of whether they are error free or not. (We can say that the receiver is blocked for M successive code words).

The transmitter A, after receiving the NAK block, resends the last M blocks that were transmitted before the NAK block. Because of (5.31), this protocol has the following important property: the first word to be received after reopening of the receiver is always the word with a detected error before blocking of the receiver. Therefore the right word order is kept, independent of a

a)

b)

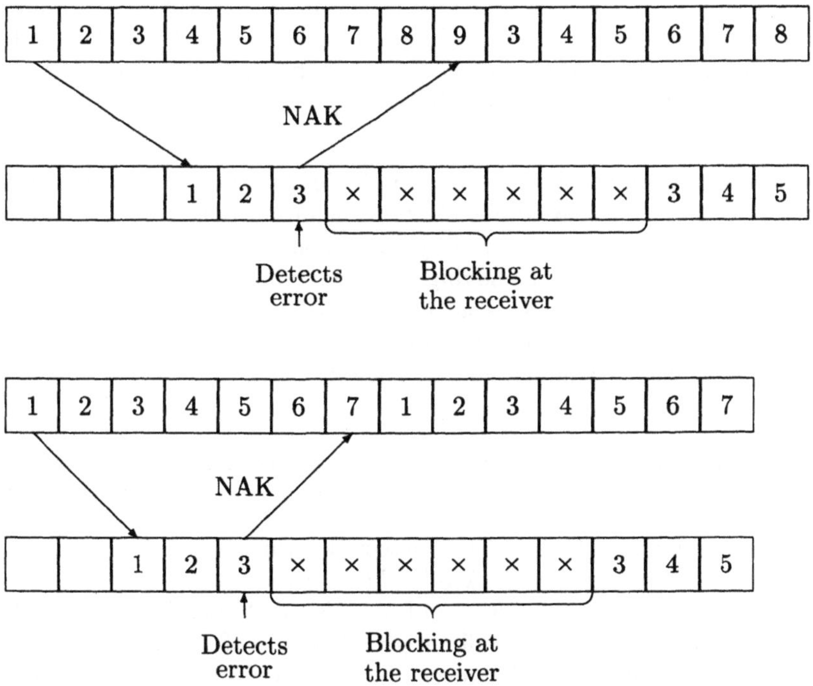

Figure 5.2 a) The time diagram of the go-back-M protocol with $M = 7$, $n_{\mathrm{rt}} = 3n$, $n_{\mathrm{s}} = n$. b) The time diagram of the go-back-M protocol with $M = 7$, $n_{\mathrm{rt}} = 2n$, $n_{\mathrm{s}} = n$.

precise knowledge of the round-trip delay in the given channel. If (5.31) is satisfied with equality, we obtain the following additional property: the first block transmitted from A to B after A receives the NAK block from B will be exactly the block where B detected an error. This situation is preferable, because it minimizes the number of retransmissions over the channel and the size of the memory at the transmitter side.

We give the time diagram of this protocol in Figure 5.2 where in the case a) we have equality in (5.31) and in the case b) inequality. So we have a trade-off between knowledge of the round-trip-delay n_{rt}, and the information transmission efficiency, and the complexity of the circuits to provide this protocol.

Remark. At the transmitting end it is in principle necessary to have a memory of unlimited size, to provide continuous work of the information source with ARQ systems. In practice, the size of the memory is of course limited, but should be so large that the probability of overflow is neglectable. (In the rare event of an overflow, the transmission must be halted temporarily; if this is impossible some information will be lost).

The go-back-M protocol can be used for duplex information transmission. Then the service signals (ACK or NAK) are transmitted both from A to B and vise versa. They can be transmitted by special service blocks of length n' or by ordinary information blocks having service symbols. But the most common is to transmit a special block as an NAK signal and to transmit nothing as an ACK signal. Then the modified *cross blocking* protocol for a duplex go-back-M ARQ system can be summarized as follows:

i) After receiving a block, without detected error, and which is not an NAK block, it is delivered to the user, and the transmitter sends the next block.

ii) After receiving a block with a detected error or an NAK block, it is discarded, the receiver is blocked for M successive code words, and the transmitter sends the NAK block and resends the M previous blocks. It is easy to see that this protocol will not give deletions of blocks after a NAK block with detected errors. However, deletions and insertions may occur with probabilities less then P_{ue} for the chosen code.

The relations (5.17) and (5.20) which were proved for the stop-and-wait protocol remains true for the go-back-M protocol, and then P_{in} and P_{de} can be bounded by P_{f} here as well as in the previous case. Moreover, if we are dealing with a channel with memory, namely the varying parameter channel, the reliability of the go-back-M protocol will be better then that given by these formulas because, as a rule, the blocking time is placed on the bad states of the channel and during the blocking time it is practically impossible to obtain undetected errors.

Let us now consider the delivery time efficiency of this protocol. Denote by Δ' and Δ'' the probabilities of the events that the given transmission of the block will be the last attempt to convey it in the directions A$\rightarrow$B and B$\rightarrow$A respectively for the simplex mode. Similarly to the determination of these values for the stop-and-wait protocol by (5.25) we get

$$\begin{aligned} \Delta' &= P_{\text{ef}}P'_{\text{ef}} + P_{\text{ue}}P'_{\text{ef}} + P_{\text{d}}P'_{\text{ue}} \ , \\ \Delta'' &= P'_{\text{ef}}P_{\text{ef}} + P'_{\text{ue}}P_{\text{ef}} + P'_{\text{d}}P_{\text{ue}} \ . \end{aligned} \tag{5.32}$$

The probability for an attempt to be the last attempt to convey a block for duplex mode with cross blocking and independent forward and return channels is given by

$$\Delta = \Delta'\Delta'', \tag{5.33}$$

and for identical channels we get

$$\Delta = \left(P_{\text{ef}}^2 + P_{\text{ue}}P_{\text{ef}} + P_{\text{d}}P_{\text{ue}}\right)^2 . \tag{5.34}$$

If we have identical channels and we can assume that both are noiseless for NAK transmission, then (5.32) and (5.34) yield

$$\Delta = \left(P_{\text{ef}} + P_{\text{ue}}\right)^2 . \tag{5.35}$$

For a completely noiseless return simplex channel we get

$$\Delta = P_{\text{ef}} + P_{\text{ue}}. \tag{5.36}$$

The average number of code blocks $\bar{S}$ which have to be transmitted over the BSC per accepted information block with the go-back-M protocol can be found by a modification of (5.27)

$$\bar{S} = \Delta \sum_{i=1}^{\infty} (M(i-1)+1)(1-\Delta)^{i-1} = 1 + \frac{M(1-\Delta)}{\Delta},$$

and the throughput efficiency of this protocol in the case of transmission of an NAK signal by a special code block will be

$$\eta = \frac{k}{n} \cdot \frac{\Delta}{\Delta + M(1-\Delta)}, \tag{5.37}$$

where Δ should be calculated by (5.32), (5.33), (5.34), or (5.35) depending on whether we have a simplex or a duplex mode and on the type of channel.

Remark. Since we use a special code block to transmit the NAC signal, the minimum value of M can be found from (5.31) as $M \geq \lceil (2n + n_{\rm rt})/n \rceil = 2 + \lceil n_{\rm rt}/n \rceil$.

On the BSC, the following lower bounds on Δ' and Δ'' will be true

$$\Delta'\Delta'' \geq P_{\rm ef} P'_{\rm ef} = \Big((1-p)(1-p')\Big)^n.$$

With identical forward and return channels this gives

$$\Delta \geq (1-p)^{4n}, \tag{5.38}$$

and, instead of (5.35), for noiseless return channel we get

$$\Delta \geq (1-p)^{2n}. \tag{5.39}$$

We can see from (5.37) that for low noise forward and return channels, when $M\left(1 - \Big((1-p)(1-p')\Big)^n\right) \ll \Big((1-p)(1-p')\Big)^n$, a blocking does not affect the TE much, and so this protocol is approaching an perfect one in this case, but for noisy channels and large values of M we can get very small values of TE. To improve this situation we can try to change the protocol or to use a combination of error detection and correction (hybrid methods). The first approach is considered in the next subsection and the second will be discussed later in this section.

Now we consider the behaviour of this protocol for channels with memory. It has been shown in 4.2 and 4.3 that for some channels with memory, but without channel transforms, and for any channel with memory when channel transforms are used, we can obtain any given value of $P_{\rm ue}$ for a moderate redundancy of the linear code. The lower bound (5.19) remains true for channels with memory and for this protocol, if the probability $P_{\rm ef}$ is calculated for the given channel with memory. Moreover, for some channels with memory (see Figure 1.1) $P_{\rm ef}$ is higher than the corresponding $P_{\rm ef}$ for the BSC and therefore there is no problem to provide a given reliability for any channel with memory using a go-back-M ARQ protocol (it is only necessary to take into consideration the fact mentioned above that formula (5.8) must be changed to the formula (5.9)).

Now we return our attention to the problem of the TE calculation for this protocol for channels with memory and in simplex mode. Note first of all that a sufficient condition for any code block transmitted over the channel to be passed on to the user and never be repeated over the channel is an error free interval of M adjacent blocks both in the forward and return channels. Under this condition it is not necessary to have any repetition of the given block as we can see from Figure 5.2, and therefore, for the mutually independent forward and return channels we obtain the following lower bound on the TE in simplex mode

$$\eta \geq \frac{k}{n} P_{\text{ef}}(nM) P'_{\text{ef}}(nM) \tag{5.40}$$

where $P_{\text{ef}}(nM)$ is determined by (5.1) for a block length nM in the forward channel and $P'_{\text{ef}}(nM)$ similarly for the return channel.

If the channels are identical, we get from (5.40)

$$\eta \geq \frac{k}{n} P^2_{\text{ef}}(nM). \tag{5.41}$$

To derive a more precise expression for TE it is necessary to specify the channel model. Having considered a number of models given in Chapter 1, we can see that the more appropriate for this case is a channel model based on the pseudo-Markov chains (FPMS), because this gives the simplest description of the block stream. We assume that for a reasonable ARQ system $P_{\text{ue}} \ll P_{\text{ef}}(n)$ and therefore $P_{\text{d}} \approx 1 - P_{\text{ef}}(n)$, so the error is undetected with probability $P_{\text{ef}}(n)$. Taking into account this fact and that there are $M - 1$ code blocks between every repetition of a given code block, we obtain the following expression for the average number of codes block $\bar{S}$ which are necessary to transmit over a forward simplex channel described by the FPMC model with a noiseless return channel

$$\bar{S} = \bar{a} \left\{ \sum_{i=1}^{\infty} (M(i-1)+1) \tilde{\tilde{A}}_1^{i-1} \right\} \tilde{\tilde{A}}_0 \bar{b}^{\text{t}} \tag{5.42}$$

where

- $\tilde{\tilde{A}}_i = \tilde{A}_i \tilde{A}^{M-1}$, $i = 0, 1$,
- $\tilde{A} = A^n$,
- $\tilde{A}_0 = A_0^n$,
- $\tilde{A}_1 = A^n - A_0^n$,
- $\bar{a}$, $\bar{b}$, A_0, A_1 are the vectors and matrices which are used for the model described in Chapter 1.

Using a well known matrix formula (Birkhoff&MacLane [3])

$$\sum_{i=0}^{\infty} \tilde{\tilde{A}}_1^i = \left(I - \tilde{\tilde{A}}_1\right)^{-1}, \tag{5.43}$$

where I is the identity matrix, and calculating a formal derivative with respect to $\tilde{\tilde{A}}_1$, we get

$$\sum_{i=0}^{\infty} i\tilde{\tilde{A}}_1^{i-1} = \sum_{i=1}^{\infty} i\tilde{\tilde{A}}_1^{i-1} = \left(I - \tilde{\tilde{A}}_1\right)^{-2}. \tag{5.44}$$

Substituting (5.43) and (5.44) into (5.42) we get

$$\bar{S} = \bar{a}\left\{M\left(I - \tilde{\tilde{A}}_1\right)^{-2} - (M-1)\left(I - \tilde{\tilde{A}}_1\right)^{-1}\right\}\tilde{\tilde{A}}_0\bar{b}^t. \tag{5.45}$$

Thus the TE of the go-back-M simplex ARQ protocol for the channel with memory described by the FPMS model and a noiseless return channel is

$$\eta = \frac{k}{n}\left\{\bar{a}\left\{M\left(I - \tilde{\tilde{A}}_1\right)^{-2} - (M-1)\left(I - \tilde{\tilde{A}}_1\right)^{-1}\right\}\tilde{\tilde{A}}_0\bar{b}^t\right\}^{-1}. \tag{5.46}$$

Example: We consider the RFC model described in Chapter 1 and take as its parameters $\alpha = 1/2$, $h^2 = 48$, and (a) $r = 0.5$ or (b) $r = 0.995$. Then the average symbol error probability will be $p = 2 \cdot 10^{-2}$ according to (1.12). Using (1.36)–(1.39) we obtain the FPMC model corresponding to the original RFC model with rank $l = 3$ and the following parameters

(a) $r = 0.5$, $\bar{a} = (1,\ 1.96 \cdot 10^{-2},\ 1.96 \cdot 10^{-2})$, $\bar{b} = (1,\ 0,\ 0)$,

$$A_0 = \begin{pmatrix} 1 & -1.03 \cdot 10^{-6} & 1.4 \cdot 10^{-2} \\ -1 & -2.55 \cdot 10^{-6} & -3.23 \cdot 10^{-2} \\ 2.74 \cdot 10^{-8} & 1 & 2.95 \cdot 10^{-1} \end{pmatrix},$$

$$A_1 = \begin{pmatrix} -5.33 \cdot 10^{-13} & 1.034 \cdot 10^{-6} & 1.72 \cdot 10^{7} \\ 1 & 2.55 \cdot 10^{-2} & 2.08 \cdot 10^{-2} \\ 3.44 \cdot 10^{-11} & -8.7 \cdot 10^{-5} & -1.45 \cdot 10^{-5} \end{pmatrix}.$$

(b) $r = 0.995$, $\bar{a} = (1,\ 1.96 \cdot 10^{-2},\ 1.96 \cdot 10^{-2})$, $\bar{b} = (1,\ 0,\ 0)$,

$$A_0 = \begin{pmatrix} 1 & -3.98 \cdot 10^{-4} & 4.15 \cdot 10^{-5} \\ -1 & -9.5 \cdot 10^{-1} & -1.43 \\ 8.98 \cdot 10^{-11} & 1.74 & 2.23 \end{pmatrix},$$

$$A_1 = \begin{pmatrix} 2.84 \cdot 10^{-14} & 3.98 \cdot 10^{-4} & 2.87 \cdot 10^{-4} \\ 1 & 9.5 \cdot 10^{-1} & 7.65 \cdot 10^{-1} \\ 0 & -7.43 \cdot 10^{-1} & -5.74 \cdot 10^{-1} \end{pmatrix}.$$

We assume that the block length $n = 128$ and $M = 3$. Using formula (5.45) we obtain $1/\bar{S} = 0.0285$ for $r = 0.5$ and $1/\bar{S} = 0.142$ for $r = 0.995$.

We observe that for the small correlation coefficient $r = 0.5$, the value of $1/\bar{S}$ almost coincides with that for the BSC without memory given by the factor in (5.37) for $\Delta = (1-p)^n$, where $p = 2 \cdot 10^{-2}$ and $n = 128$.

On the other hand, for the large correlation coefficient $r = 0.995$, the value of $1/\bar{S}$ is more signicant. This illustrates our previous statement about the positive effect of memory on the TE of feedback systems.

Let us try to increase TE using combined error detection and correction. It has already been mentioned that if we use all the detection capability for correction, we will be able to bring TE closer to the capacity of the channel. And since feedback does not increase the capacity of a channel without memory, it may appear that error detection is useless. However, this approach usually requires a very complex decoding procedure. Therefore, we try to correct only a moderate number of errors and will do this, not to increase the reliability of the information transmission, but only to increase the TE.

If we use combined error detection and correction by some linear $[n, k]$ code V, we get two quantities which describe this situation: $P_{\mathrm{ue}}^{(t)}(V)$, the probability of undetected error when V is used to correct at most t errors, and $P_{\mathrm{c}}^{(t)}(V)$, the probability of correct decoding. The value $P_{\mathrm{c}}^{(t)}(V)$ is given by the following expression for channels with and without memory respectively:

$$P_{\mathrm{c}}^{(t)}(V) = 1 - P(n, t+1), \tag{5.47}$$

where $P(n, m)$ is the the multiplicity function of the given channel, and

$$P_{\mathrm{c}}^{(t)}(V) = \sum_{i=0}^{t} \binom{n}{i} p^i (1-p)^{n-i}. \tag{5.48}$$

For the go-back-M protocol we can find P_{f} and TE by changing P_{ef} to $P_{\mathrm{c}}^{(t)}(V)$ and P_{ue} to $P_{\mathrm{ue}}^{(t)}(V)$ everywhere. In turn these values can be found or estimated using the relations or bounds proved in Chapters 3 and 4.

Example. We consider a go-back-M simplex protocol with $M = 3$, for a forward BSC having symbol error probability $p = 10^{-2}$ and a noiseless return channel. Suppose that the required reliability is $P_{\mathrm{eq}} = 10^{-6}$ for every symbol error probability in the interval $[0, 1/2]$. To satisfy this condition requires a t-good code or stochastic transform. Choose a good $[128, 113]$ code for error detection only. According to (3.1) or (4.31), (5.11), (5.21), (5.36), and (5.37), this provides a throughput efficiency equal to 0.1. If we take the same $[128, 113]$ code it is possible to add $r' = 8$ check symbols to obtain a 1-good $[136, 113]$ code. Using (3.19), (5.11), (5.21), (5.36), (5.37), and (5.48), we get that this code and a combined algorithm for error detection and correction give almost the same reliability $P_{\mathrm{eq}} \approx 10^{-6}$ and TE $\eta \approx 0.28$. We can conclude that a combined algorithm allows us to improve the TE, especially for bad channels, but it is only slightly better than a pure error detection algorithm for good channels. Therefore, we need more efficient methods to incorporate correction and detection, in particular methods which combines code blocks with detected error with retransmitted copies, or some iterative algorithms, but we will discuss these approaches in the next subsections.

The Selective Repeat ARQ Protocols

The go-back-M protocol is a system with "mutual responsibility", since $M - 1$ possibly good received code blocks are discarded because of detected errors in some preceding block. To avoid this fallacious strategy, one can repeat only the single block with detected errors. However, this approach results in the necessity of solving the following problems:

1. To determine at the transmitting end the previous transmitted code block to which an NAK block relates.

2. To determine at the receiving end the position of a retransmitted block. (As a rule it is necessary to deliver the code blocks in correct order to the user).

The first problem can be solved, in principle, if we append a special address group to every code block to indicate their order. This also requires a retransmission buffer. The use of addresses requires additional service symbols in the code blocks and therefore decreases the code rate and TE. (We can avoid the use of addresses if we know the round-trip-delay in the given channel exactly. However, this is rather complicated to implement).

Note that a receiver buffer of any finite size can overflow because there is a nonzero probability that any code block may be retransmitted any number of times. We can increase the size of the buffer to bring this probability closer to zero, but this is possible only in the case when we increase the range of the sequence numbers (addresses) and this in turn decreases the TE. As a rule, the block addresses are used cyclically from 1 to N (i.e. a new code word following a code word with address number N is given address number 1).

We can see that there exists many problems to realize a selective repeat strategy and there are therefore many different protocols to implement it under different conditions. We consider the main versions, but without some details which can be found in the references given at the end of the chapter.

A Perfect Selective Repeat Protocol

We assume here that the buffer size can be chosen unlimited and we do not use addresses for information transmission. In this case we obtain the same results as for the stop-and-wait protocol, but with $n_{\text{it}} = 0$, namely

$$P_{\text{f}} = \frac{P_{\text{ue}}}{P_{\text{ue}} + P_{\text{ef}}}. \tag{5.49}$$

For channels without memory

$$P_{\text{eq}} \approx \frac{P_{\text{f}}}{k}. \tag{5.50}$$

The average symbol error probability after using this protocol is approximately

$$P_{\text{av}} = \frac{d}{n} P_{\text{f}}, \tag{5.51}$$

where d is the minimum code distance for the given code. For a simplex protocol the throughput efficiency will be

$$\eta = \frac{k}{n\bar{S}} = \frac{k}{n}\Delta, \tag{5.52}$$

where Δ is calculated by (5.25) or is lower bounded by (5.28).

For a duplex protocol which uses service symbols for ACK or NAK transmission in every code block we have

$$\eta = \frac{k-1}{n}\Delta. \tag{5.53}$$

Remark. We have collected here a list of formulas for estimation of the perfect strategy efficiency (and even repeated some results from previous subsections) for future use as bounds for different real protocols and in particular for the code and protocol optimization procedure considered in Chapter 6.

If we use addresses numbered modulo N, all the formulas above will be true if we substitute k with $k' = k - m$, where $m = \lceil \log_2 N \rceil$.

When combined error correction and detection is used with a perfect SR protocol for the BSC it is necessary to change P_{ue} to $P_{\mathrm{ue}}^{(t)}(V)$ in all formulas of Chapter 3 and P_{c} to (5.47), where t is the maximal number of errors corrected.

For the channels with memory described by FPMC we can use formula (5.46) with $M = 1$ for simplex systems and $M = 1$ (in which case k is changed to $k - 1$) or $M = 2$ for duplex systems.

Selective Repeat Protocol with Finite Receiver Buffer

This protocol has been completely described by Lin and Costello [11]. Here we only give the key points of this strategy and the main formulas for TE calculation. (the details about transmission, retransmission, and error recovery procedures can be found in [11]).

Consider this protocol with a finite receiver buffer of size N and simplex systems. Let N' be the number of code blocks that can be transmitted during one round-trip delay period and take $N = N'$. (This is obviously the minimum possible value of the buffer size). The range of the addresses is the set $\{1, 2, \ldots, 3N\}$. The receiver will then be able to distinguish whether a received vector with address q is a new vector or a vector that has been accepted and delivered.

The transmission and retransmission procedure is to send a new block when the received ACK signal is received over the return channel and otherwise resend the old block with the address contained in the NAK signal received.

The receiver's operations are the following: to detect errors, to deliver blocks without detected errors to the user in correct order, to release from the buffer the code words delivered to the user, and finally, to reject code words and send NAK signals after a buffer overflow.

We can see that detection of a second error in one code word results in rejection of all received blocks up to the point when the erroneous word is received without detectable errors and then the buffer is released. Simultaneously a NAK signal is transmitted in every rejected block whether errors are detected there or not. This strategy allows us to avoid loss of information after receiver buffer overflow for any given buffer size. But the larger the buffer, the shorter the rejection periods will be, and so the higher the throughput efficiency will be. It has been shown by Lin and Costello [11] that, for the BSC with symbol error probability p, a noiseless return channel, and a buffer of size N equal to the round-trip-delay, TE is lower bounded as follows.

$$\eta \geq \frac{\lambda_0}{\lambda_0 + \lambda_1 + \lambda_2 N} \cdot \frac{k}{n} \tag{5.54}$$

where

$$\begin{aligned}
\lambda_0 &= \frac{\Phi_1}{1-\Phi_2}\left(1 - \Phi_1 \Phi_2^{N-1}\right),\\
\lambda_1 &= P_c\left\{\Phi_0^{N-2} + (1-P_c)\,\Phi_1^{N-2} + (1-P_c)^2\,\Phi_2^{N-2}\right\},\\
\lambda_2 &= 3 - P_c^2\Phi_0^{N-2} - \Phi_0^2\Phi_1^{N-2} - \Phi_1\Phi_2^{N-2},\\
\Phi_0 &= 1-(1-P_c)^2,\ \Phi_1 = 1-(1-P_c)^3,\ \Phi_2 = 1-(1-P_c)^4,\\
P_c &= (1-p)^n.
\end{aligned}$$

(The fact that m bits are used for address transmission is ignored here, because as a rule $m \ll k$).

This protocol is very close to a *mixed mode* ARQ, namely the *selective repeat plus go-back-M*, which is also described by Lin and Costello [11]. In this case the transmitter remains in a selective repeat mode, until ν retransmissions are made, but the transmitter fails to receive ACK for some code word. After more than ν retransmissions, the system *switches* to the go-back-M retransmission mode. The transmitter stops to send new words and resends $M-1$ succeeding words that were transmitted following the ν-th SR retransmission attempt for the unlucky word. The transmitter stays in this mode until this word is positively acknowledged. At the receiver after ν transmissions, the subsequent $M-1$ received words are discarded regardless of whether they contain detected errors or not.

To calculate the TE for this mixed mode with a noiseless return channel and given parameter ν, when the receiver buffer must be able to store $\nu(M-1)+1$ code words, we consider the signal-flow graph corresponding to this protocol (see Appendix 5.6 for details). This graph is plotted in Figure 5.3.

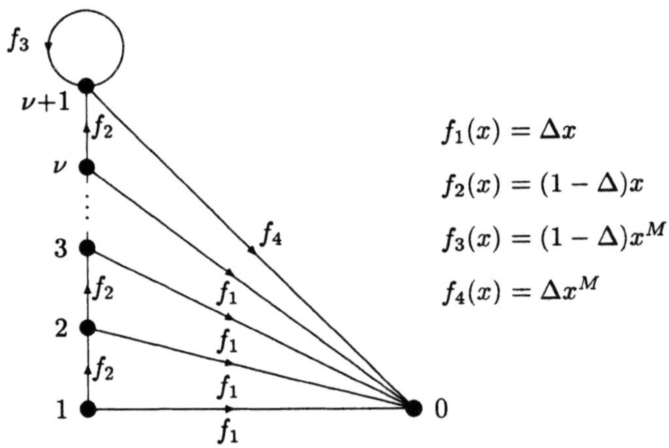

Figure 5.3 The graph of the single information word transmission for SR-GBM mode with given value ν.

Node 1 corresponds to the first attempt to transmit a code word. The nodes $2, 3, \ldots, \nu$ correspond to the second, third, ..., ν-th attempts to convey it after receiving the NAK signals. The $\nu+1$-th node corresponds to the transition from the SR to the GBM mode. All the branches of this graph except two, namely $(\nu+1, 0)$ and the loop $(\nu+1, \nu+1)$, require one code block for transmission over the channel; whereas the loop requires M code blocks. Assuming that all the transmissions through branches which lead to node 0 have probabilities $\Delta = P_{\text{ef}} + P_{\text{ue}}$, and all other branches have probabilities $1-\Delta$, we get the generating branch functions shown in Figure 5.3.

Using the transformation rules given in Appendix 5.6, we obtain the following generating function to get from node 1 to node 0

$$f(x) = \frac{f_2(x)^{\nu+1} f_4(x) + f_1(x)(1-f_3(x)) \sum_{i=0}^{\nu} f_2(x)^i}{1-f_3(x)}. \tag{5.55}$$

Putting the branch generating functions into (5.55) and differentiating this relation with respect to x we obtain, after the substitution $x = 1$, the following expression for the average number of code blocks necessary to be transmitted over BSC channel for final acceptance of one information block for the simplex mode

$$\bar{S} = \frac{1 + (M-1)(1-\Delta)^{\nu+1}}{\Delta}$$

and thus the throughput efficiency of the mixed mode of ARQ (SR+GBM) will be the following

$$\eta = \frac{k}{n} \cdot \frac{\Delta}{1 + (M-1)(1-\Delta)^{\nu+1}}. \tag{5.56}$$

(Since, as a rule, the number of address bits is less than k, we ignore their effect).

Example. Take $n = 1024$, $k = 1000$, $M = 512$. Then for the BSC with symbol error probability $p = 10^{-4}$ in the forward channel and a noiseless return channel, we get TE for the perfect SR and GBM+SR given in Table 5.1 for different values of ν. Also given (in the denominator) is the size of the receiver buffer (in Kbytes) which is necessary to provide GBM+SR ARQ with given ν is shown.

Table 5.1 The throughput efficiency for a [1024,1000] code, BSC with $p = 10^{-4}$, $M = 512$, and different ARQ protocols.

Type of protocol	TE	TE/size of buffer		
		$\nu = 1$	$\nu = 2$	$\nu = 5$
Perfect selective repeat	0.88	-	-	-
Go-back-M	0.017	-	-	-
Go-back-M + selective repeat		0.15/64	0.6/128	0.88/320

From Table 5.1 we can see that for a given code and a channel using the GBM protocol, a very low TE is provided, but a mixed GBM+SR protocol with $\nu = 5$ gives practically the same results as the perfect SR strategy. To decrease the size of the buffer, one can limit the value of ν to $\nu = 2$. This provides a TE which is 68% of the maximum possible value.

ARQ Protocol With Address Confirmation

The ARQ protocols with address retransmission on request are superior to address confirmation protocols only in the case when a simplex information system or a duplex system with special NAK signals is used. (In this case, the channel is used more rarely for the transmission of service signals, because the NAK signal for a reasonable system is less probable than the ACK). But, if we use a duplex system and special service symbols in each code block to convey NAK or ACK signals, this advantage can be lost. On the other hand, we will show that the ARQ protocol with address confirmation can be used to reduce a real return channel to a noiseless channel. The main idea of this approach has been proposed by Amosov and Aleksandrov [26] and it exploits an opportunity to confirm the addresses of some code blocks received earlier without detected errors.

For this protocol and a channel without memory, when the transmitter and receiver buffers have unlimited sizes, the following lower bound for TE is true

$$\eta \geq (1 - P_{\mathrm{d}})(1 - (P'_{\mathrm{d}})^{s})\frac{k'}{n}, \tag{5.57}$$

where

- P_{d} and P'_{d} are the probabilities of detected errors in the forward and return channels respectively,
- $k' = k - sm$,
- $m = \lceil \log(3M) \rceil$ is the number of bits which are used to transmit addresses (M is the round trip delay in terms of the number of code blocks),
- s is a parameter that can be optimized.

The relation (5.57) shows that for channels with moderate values of the round trip delay we get practically a noiseless return channel after using this protocol and as a rule it is sufficient to take $2 \leq s \leq 5$ to reach the highest TE.

For block lengths of more than 100 bits, the effect of the address bits on the TE can be neglected.

It has been shown by Amosov and Aleksandrov [26] that the TE for this protocol, with finite size of the buffers and for channels without memory, has the following bounds

$$1 - (1 - \tilde{p})^{(H-1)/M} \leq \eta \leq \tilde{\tilde{p}}\frac{k'}{n}, \tag{5.58}$$

where

- $\tilde{p} = (1 - P_d)(1 - (P'_d)^s)$,
- $\tilde{\tilde{p}} = \tilde{p} + (1 - P_d)(1 - P'_d) P_d (P'_d)^s / (1 - P_d P'_d)$,
- H is the size (in number of code blocks) of the receiver buffer,
- M is the round trip delay (in number of blocks),
- P_d, P'_d, k, and k' have the same meaning as in (5.57).

(A computer simulation of this protocol showed that it is efficient, also on High-Frequency channels).

ARQ Selective Repeat Protocol for Session Transmission Systems

Let us consider a session transmission where we have to convey a rather small number M of information blocks. One can then consider the following ARQ protocol. We transmit the M blocks, all of them by individual addresses. At the receiving end, possible errors are detected and the addresses of the blocks with detected errors are transmitted over the return channel. Next the transmitter resends the blocks with these addresses and the receiver performs error detection again. The whole procedure is repeated until all the blocks are received without detected errors. Then all the blocks are delivered to the user in correct order.

It is natural to choose as a measures of reliability and efficiency of this system two characteristics: the probability $P_c(M)$ of receiving all M blocks without any errors and the probability distribution function of the transmission time for the whole session. The function $P(M, S)$ can be defined as the probability of transmitting M information blocks with this protocol by transmitting at most S blocks over the channel, where $S \geq M$.

We have restricted ourselves to the case of a noiseless and instant return channel. (This is not a very strong restriction for a simplex system).

For the BSC with symbol error probability p we get the following relation for the probability of receiving correctly the whole session consisting of S blocks

$$P_c(M) = (1 - P_f)^M, \tag{5.59}$$

where

- $P_f = P_{ue}(C, p) / (P_{ue}(C, p) + (1 - p)^n)$,

- $P_{ue}(C,p)$ is the probability of undetected error for a given code C and symbol error probability p.

For channels K with memory we have an upper bound

$$P_c(M) \leq MP_f, \tag{5.60}$$

where

$$P_f = \frac{P_{ue}(C,K)}{P_{ue}(C,K) + P_{ef}(n)}.$$

When some hybrid method is used and the code corrects at most t errors and detects other errors, it is necessary to change P_f in (5.59) and (5.60) to

$$P_f = \frac{P_{ue}^{(t)}(C,K)}{P_{ue}^{(t)}(C,K) + 1 - P(n,t+1)}, \tag{5.61}$$

where

- $P_{ue}^{(t)}(C,K)$ is the probability of undetected error for a given code C and channel K when correction of at most t errors is used in combination with error detection,
- $P(n,m)$ is the multiplicity function for the channel K.

(All the values contained in the formulas (5.59)–(5.61) can be found for different codes and channels in Chapters 3 and 4).

The graph of this session protocol is shown in Figure 5.4.

The node I is the initial state, when the transmission of M code blocks is begun. The states $i \in \{0, 1, \ldots, M\}$ corresponds to receiving i of the M code blocks with detected errors. The transition from state I to any state i requires transmission of M blocks and the transition from state i to state j requires transmission of i blocks. We need to derive the generating function between state I and state 0. The probability to get from state I to state j is

$$P(j/I) = \binom{M}{j} P_d^j (1 - P_d)^{M-j} \tag{5.62}$$

and the probability to get from state i to state j is

$$P(j/i) = \binom{i}{j} P_d^j (1 - P_d)^{i-j}, \tag{5.63}$$

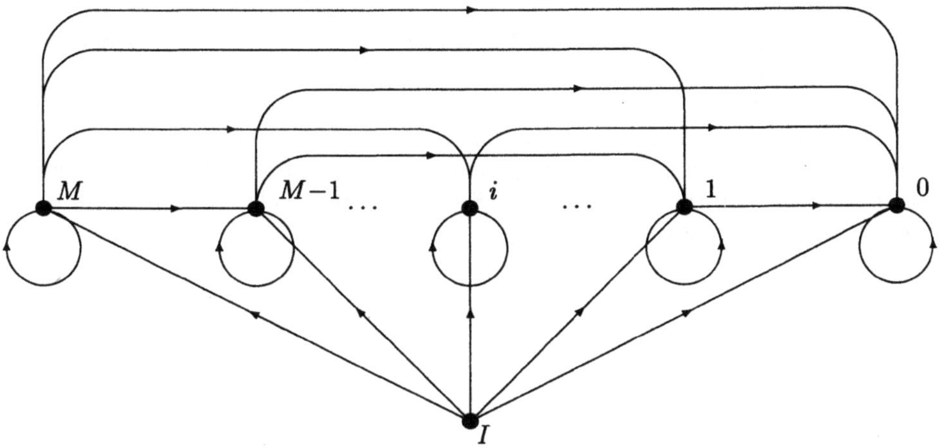

Figure 5.4 The graph of the session transmission ARQ protocol.

where P_d as usual is the probability of a block with detected error.

Thus the generating functions of the branches (I, j) will be

$$f_{Ij}(x) = P(j/I)x^M, \text{ for } 0 \leq j \leq M, \tag{5.64}$$

and the generating function of the branches (i, j) will be

$$f_{ij}(x) = P(j/i)x^i, \text{ for } 0 \leq j \leq i \leq M. \tag{5.65}$$

Using the rules for reduction of the flow graphs given in Appendix 5.6, we can find the generating function between the states I and 0, and its coefficient for x^s is the probability to reach 0 from I through s "steps". This is just the distribution function of the transmission time for the whole session. If we find $f'_{I0}(1)$, this gives us by (5.103) the average number of code blocks required to transmit the whole session and thus makes it possible to calculate the average time for the transmission session.

For the particular case $M = 2$ we get the following generating function:

$$f_{I0}(x) = \frac{x^2 (1 - P_d)^2}{(1 - P_d x)^2}. \tag{5.66}$$

We can now expand (5.66) into a power series in x, and the coefficients gives the distribution function.

In general it is not so easy to find the generating function $f_{I0}(x)$ using the graph shown in Figure 5.4 and the branch generating functions given by (5.62)–(5.65). But fortunately there exists a very simple expression for the average number of code blocks $\bar{S}$ needed to transmit the whole session consisting of M information blocks. It follows from a well known theorem in probability theory: the expectation of a sum equals the sum of the expectations. This theorem in combination with the expression for $\bar{S}$ with a perfect selective repeat protocol which is used here gives us

$$\bar{S} = \frac{M}{1 - P_{\mathrm{d}}}. \tag{5.67}$$

We can check that for $f_{I0}(x)$ given by (5.66), $f'_{I0}(1) = 2/(1-P_{\mathrm{d}})$, this coincides with (5.67) for $M = 2$.

This can be generalized, in the case of a noisy return channel, to

$$\bar{S} = \frac{M}{\Delta}, \tag{5.68}$$

where Δ is given by (5.25).

Unfortunately, there is no simple expression for the distribution function of this protocol, but we can give a lower bound in some cases. If receiving one information block without detected errors requires at most S code blocks sent, them a whole session of M blocks requires at most MS blocks sent. Therefore, we obtain

$$P(M, x = MS) \geq \left\{(1 - P_{\mathrm{d}}) \sum_{i=0}^{S-1} P_{\mathrm{d}}^{i}\right\}^{M} = \left(1 - P_{\mathrm{d}}^{S}\right)^{M}, \tag{5.69}$$

for $S = 1, 2, \ldots$. This bound can be generalized, for a noisy channel, to

$$P(M, x = MS) \geq \left(1 - (1 - \Delta)^{S}\right)^{M}, \tag{5.70}$$

where Δ is determined by (5.25).

Iterative and Concatenated Protocols

To increase the throughput efficiency for ARQ systems one can combine error correction and detection, but this requires larger complexity of the hardware or software for the decoding. Another approach consists in joint processing of the received blocks to improve reliability for a given redundancy. There are several such methods which use both the information contained in blocks with detected errors and special coding with correlation between blocks. The first group of methods are called ***hybrid ARQ systems of type II*** and the second *iterative* or *concateneted* ARQ systems. We start with the first group. The idea can be realized with a joint selective repeat ARQ protocol, and we will assume here that the return channel is noiseless.

Let us consider some systematic binary linear $[n,k]$ code V_1 which is used for error detection only. Having received some block $\tilde{v}$ with detected errors, we would like to transmit over the channel some additional information to be used with the corrupted block $\tilde{v}$ to correct the errors and provide the user with k information symbols with a given reliability. How can we construct this additional information? To avoid some trivial decisions, we introduce some natural restrictions. First, we assume that the second block has the same length n. We further assume that we have to detect errors after the second transmission also, otherwise we will not manage to provide a given reliability for a bad channel. This leads us to choose the same code for the second transmission. Thus there is only one problem — how to correlate the information symbols of the second block with the information symbols of the first block $\tilde{v}$. Assuming that the check symbols of the first block are not used for error correction after the second block is received, we get a very natural decision, namely to take as information symbols of the second block the check symbols of some systematic $[2k,k]$ code V_2, where the information symbols coincide with the information symbols of the first block. Thus we have

$$\begin{array}{ll} v_1 = (u, f(u)), & \tilde{v}_1 = v_1 + e_1, \\ v_2 = (q(u), f(q(u))), & \tilde{v}_2 = v_2 + e_2, \end{array} \qquad (5.71)$$

where the addition is modulo 2 as usual, and

- f is the function used to form the check symbols of the code V_1,
- q is the function used to form the check symbols of the code V_2,
- v_1 is the first received block, its error pattern is e_1
- v_2 is the second received block, its error pattern is e_2.

We need also to choose the code V_2 such that q is *invertible*, i.e. we can determine the information symbols from the check symbols. This is no problem. For instance, Lin and Costello [11] show that if V_0 is a cyclic $[n', n'-k]$ code with $n' \geq 2k$, then the shortened $[2k,k]$ code V_2 has this property. Clearly,

V_2 has at least the same error correcting capability as V_0. If $g(z)$ is the generator polynomial for the code V_0, and $s(z)$ is the polynomial representation of the parity-check block $q(u)$, then the polynomial representation $u(z)$ of the information can be determined as the residue of $s(z)z^l$ modulo $g(z)$.

To guarantee high reliability for this procedure, we need to use the code V_2 not only for error correction, but for correction combined with detection. Let us assume that V_2 is constructed with minimum distance d, and we will correct at most t errors (where $2t < d$). The probability of undetected error $P_{\text{ue}}^{(t)}(V_2, K)$ for a given channel can be calculated by the formulas given in Chapers 3 and 4. It is possible to choose the code V_2 such that $P_{\text{ue}}^{(t)}(V_2, K) \leq P_{\text{ue}}(V_1, K)$. If the second transmission has been unsuccessful, u is discarded, but the erroneous block $q(u)$ is stored. After an unsuccessful third transmission $q(u)$ is discared and u is stored and so on; the retransmissions are alternating repetitions of the parity code vector v_2 and the information vector v_1. The receiver stores the received message u and parity block $q(u)$ alternately. This procedure continues until u is recovered (possibly with undetected errors). If V_2 has the same error detecting capability as V_1, the reliability of this protocol is the same as the selective repeat protocol decribed in the previous section.

To find the TE for this protocol we consider its state graph which is shown in Figure 5.5 for unlimited buffer size and noiseless return channel.

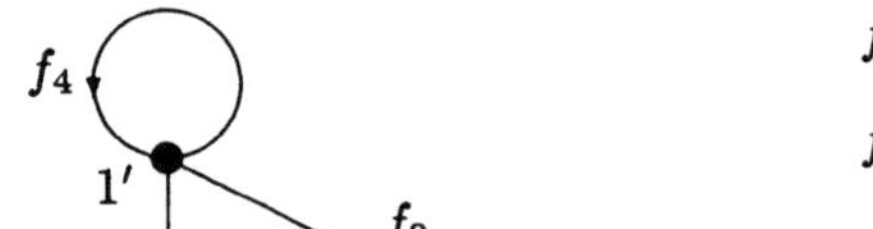

$$f_1(x) = (1 - P_{\text{d}})\,x$$

$$f_2(x) = P_{\text{d}} x$$

$$f_3(x) = (1 - P'_{\text{d}})\,x$$

$$f_4(x) = P'_{\text{d}} x$$

Figure 5.5 The graph of the hybrid type II protocol with an unlimited buffer size and noiseless return channel.

Node 1 corresponds to the first transmission of the first information block over the channel as a v_1 block, node 2 is the first transmission of the second information block, and node $1'$ corresponds to the retransmission of the information blocks.

For odd retransmissions we send a v_2 block, and for even retransmissions we send a v_1 block; the probabilities to get from $1'$ to 2 will be the same, namely it will be the probability that either code V_1 does not detect errors, or that it detects errors but code V_2 does not detect errors after error correction. The following approximate expression can be derived for this probability on the BSC,

$$1 - P'_{\mathrm{d}} \approx (1-p)^n + (1-(1-p)^n) \sum_{i=0}^{t} \sum_{j=0}^{t-i} P_i P_j, \qquad (5.72)$$

where, letting $r = n - k$,

$$P_0 = \frac{(1-(1-p)^r)\,(1-p)^k}{1-(1-p)^n}, \qquad (5.73)$$

$$P_i = \frac{\binom{k}{i} p^i (1-p)^{k-i}}{1-(1-p)^n} \quad \text{for } i \neq 0. \qquad (5.74)$$

The transmission probability $1 - P_{\mathrm{d}}$ to get from 1 to 2 is $(1-p)^{2k} + P_{\mathrm{ue}}(V_1)$. The branch generating functions are shown in the figure. Using the graph transformation method given in Appendix 5.6, we obtain the following generating function between nodes 1 and 2,

$$f_{12}(x) = f_1(x) + \frac{f_2(x) f_3(x)}{1 - f_4(x)}. \qquad (5.75)$$

The average number of blocks $\bar{S}$ which is necessary to transmit over the BSC with symbol error probability p to receive one information block is

$$\bar{S} = \left. \frac{(f_{12}(x))'}{m} \right|_{x=1} = \frac{1 + P_{\mathrm{d}} + {P_{\mathrm{d}}}^2}{1 - {P_{\mathrm{d}}}^2 P'_{\mathrm{d}}}. \qquad (5.76)$$

Example: Let us take a [63, 45] BCH code V_1 and some [90, 45] code V_2, where we correct at most 5 errors. The results of the TE calculations for the SR type II protocol when $p = 0.01$, for different values of t, are shown in Table 5.2. The TE for a pure SR protocol (without V_2 and without error correction) is shown in the last row of the table.

Table 5.2 The throughput efficiency for the SR type II protocol with V_1 a [63,45] BCH code, V_2 a [90,45] code, $p = 0.01$, and $1 \le t \le 5$.

	$t=1$	$t=2$	$t=3$	$t=4$	$t=5$
SR type II	0.393	0.412	0.421	0.422	0.423
pure SR	0.37				

We can conclude from these results that TE for the SR hybrid type II protocol is only slightly greater than TE for an ordinary SR protocol in this case. However, the situation may be quite different for more noisy channels, where the SR hybrid type II protocol is significantly better than the SR without error correction.

For a full description, let us provide the lower bound of $\bar{S}$ which have been given by Lin and Costello [11] for buffer size N:

$$\bar{S} \geq \frac{\delta_0}{\delta_0 + \delta_1 + \delta_2 N}, \tag{5.77}$$

where

- $\delta_0 = \frac{\theta_2}{1-\theta_1}\left(1 - \theta_2\theta_1^{N-1}\right)$,
- $\delta_1 = \bar{P}_{\mathrm{d}} Q_1 \theta_0^{N-2} + \bar{P}_{\mathrm{d}}\left(1 - Q_1\right) Q_2 \theta_1^{N-2}$,
- $\delta_2 = 2 - \bar{P}_{\mathrm{d}} Q_1 \theta_0^{N-2} - \theta_0\theta_2\theta_1^{N-2}$,
- $\theta_0 = 1 - P_{\mathrm{d}}\left(1 - Q_1\right)$,
- $\theta_1 = 1 - P_{\mathrm{d}}\left(1 - Q_1\right)\left(1 - Q_2\right)$,
- $\theta_2 = Q_1 + \left(1 - Q_1\right) Q_2$,
- $Q_1 = \bar{P}_{\mathrm{d}} + \frac{1}{P_{\mathrm{d}}}\Big\{\sum_{i=0}^{t} \binom{2k}{i} p^i (1-p)^{2k-i} + (1-p)^{2n}$ $\quad -2(1-p)^n \left[(1-p)^k + \sum_{i=1}^{t} \binom{k}{i} p^i (1-p)^{k-i}\right]\Big\}$,
- $Q_2 \geq \bar{P}_{\mathrm{d}} + \frac{1}{P_{\mathrm{d}}(1-Q_1)}\left\{\sum_{i=1}^{t-1} \Delta_i S_{t-i}\left(1 - \Delta_0 - S_{t-i}\right)\right\}$,
- $\Delta_i = \binom{k}{i} p^i (1-p)^{k-i}$,
- $S_j = \sum_{i=1}^{j} \Delta_i$,

- $\bar{P}_{\rm d} = 1 - P_{\rm d}$.

The type II hybrid ARQ scheme has two main drawbacks. The first is that it uses a non-optimal choice for the n redundant symbols in the retransmissions and a non-optimal decoding algorithm for the $[2n, k]$ code which is formed after two transmissions of the n-blocks. The second is that this algorithm employs only two transmissions to correct errors. It is difficult to overcome the first problem. The reason for using ARQ schemes in the first place is that we want to avoid complex forward error correction. One approach to solve this problem has been proposed by Wicker and Bartz [202], [203]. They propose to use the hybrid ARQ protocol described above, but choose the code words v_1 and v_2 in a different way. In [202], these words are taken in a punctured MDS code such that $(v_1|v_2)$ is a code word in an MDS code of rate less than $1/2$. The use of MDS codes, in particular RS codes, maximizes the minimum distance and simplifies the calculation of $P_{\rm ue}$ since the weight distribution of MDS codes is known. In [203], they choose $v_1 = v_2$ from the $R(1,m)$ Reed-Muller code. Then $(v_1|v_1) \in R(1, m+1)$ and errors can be detected and corrected using an efficient maximum likelihood decoding algorithm (The "Green machine").

Another approach to solve the problem has been proposed by Lin and Gun [139]. Its main feature is to divide each transmission block into small subblocks, so that the combination of the first and the second transmissions results in a code word of a concatenated code. Let V be a binary linear $[N, K]$ code and U a systematic binary linear $[2k, k]$ code where $k = N/m$ for some integer m. Let $p(v')$ be the check block for U corresponding to an information block v'. Any code word $v \in V$ can be written as a combination of m subblocks of length k, i.e. $v = (v_1, v_2, \ldots, v_m)$. The structure of the $[2N, K]$ concatenated code is given in Figure 5.6.

First a block $v = (v_1, v_2, \ldots, v_m) \in V$ is transmitted to the receiver. The receiver can do both correction and detection of errors. If errors are detected, retransmission is requested. In the second transmission the block $(p(v_1), p(v_2), \ldots, p(v_m))$ is sent to the receiver. The decoder of U corrects and detects errors in each "frame" (subblock $(v_i, p(v_i))$). If any frame has detected errors, then further retransmission is requested. Otherwise it is checked if the decoded block $(\tilde{v}_1, \tilde{v}_2, \ldots, \tilde{v}_m)$ belongs to V. If not, a new retransmission is requested. This is the end of the cycle, and we begin to transmit block v again and discard the two N-tuples received previously.

The main advantage of this algorithm is the possibility to provide a detailed performance analysis for such a scheme over the BSC. (Formulas for reliability

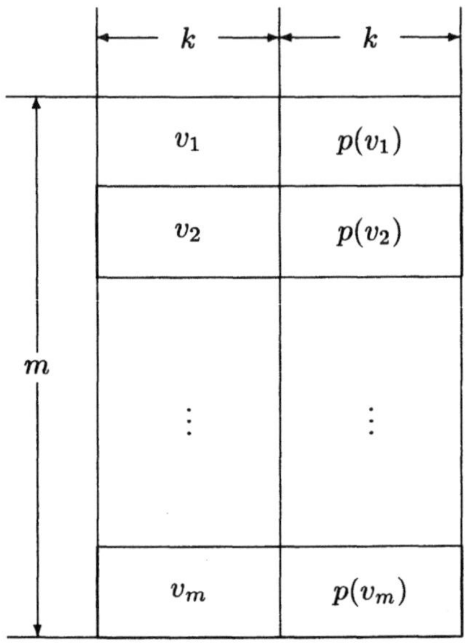

Figure 5.6 The structure of the $[2N, K]$ concatenated code.

and TE can be found in [139]). However, this approach can not be generalized to more than 2 retransmissions to improve the TE. A more general approach has been proposed by Krishna and Morgera [123]. The key point of this algorithm is to design an $[mn, n]$ code V_m which has the following properties:

1. every subblock $(a_{ni+1}, a_{ni+2}, \ldots, a_{ni+n})$ $(0 \leq i \leq m-1)$ of a code word $(a_1, a_2, \ldots, a_{mn})$ is invertible in the sence that the information n-block can be uniquely recovered from the knowledge of this subblock;

2. the encoder-decoder configuration is the same for all the codes V_i $i = 2, 3, \ldots, m$ (it does not change as the lengths of the codes increases).

This class of linear codes was discovered by Krishna and Morgera [123] (and they are henceforth refered to as KM codes). Having any $[mn, n]$ KM code V_m and some high rate $[n, k]$ code V_0 one can propose the following ARQ protocol. First the information k-block is coded into a code word in V_0. This code word is coded by V_m as $(c_1, c_2, \ldots c_m)$ where each c_i is an n-tuple. In the first transmission we send the code word c_1 and try to detect errors at the receiving side. If there are detected errors in this word a retransmission is requested. After transmission of the next subblock c_2 we try to invert it and check for errors by code V_0. If there are errors in this block we use the $[2n, n]$

code V_2 to correct errors and then we check for errors by code V_0. We continue this procedure up to the V_m code, and if errors are detected at the last stage, the receiver discards all check subblocks of the V_m code and the transmitter starts a new cycle. This protocol is called a ***generalized hybrid ARQ protocol*** (GH-ARQ). Since the redundancy of the code sequence V_i, $i = 2, 3, \ldots, m$ is increasing monotonically, more retransmissions increases the error correcting capability of the code. Therefore, the probability to have many retransmissions will be less for the GH-ARQ then for a pure type-II hybrid ARQ scheme. The graph for the GH-ARQ protocol can be constructed in a similar way as for the cases above. It is obvious that the main problem connected with GH-ARQ protocols is to construct a family of codes with properties 1 and 2. (The details of the design of KM codes can be found in [123]).

It is very important to emphasize that the GH-ARQ protocol is appropriate for channels with bursts of errors if their lengths are at most n, because due to property (1), we can recover the information block, having received any error free subblock of length n.

It is well known from Shannon's theorem [177] that for the BSC there exists a sequence of binary block codes with a given code rate R that has the following ***strong property***:

$$P(\text{ false decoding }) \to 0, \text{ when the block length } \to \infty \text{ if}$$

$$0 < R < C_{\mathrm{BSC}} = 1 + p \log p + (1-p) \log(1-p), \tag{5.78}$$

where C_{BSC} is the capacity of the BSC_p.

As a rule, for error detecting codes the throughput efficiency of an ARQ systems goes to 0 when the block length increases. Therefore, it is not obvious that there exists a sequence of linear codes with error detection only and an ARQ protocol which provides a similar property for some code rate $R > 0$.

Let us consider the following iterative protocol suggested by Jacobs [87] that satisfies (5.78) for some $R < R_0 < C_{\mathrm{BSC}}$.

We consider the product of two single parity check codes as illustrated in Figure 5.7. First we transmit and (if necessary) retransmit the first row until it is received without detected errors. The second row is transmitted the same way, and so on up til the last row. If all the columns now have even parity, the matrix is accepted and the $K_1 K_2$ symbols are delivered to the user; otherwise the whole matrix is discarded and the process starts all over. The extension of this 2 stage procedure to an iterative code with M stages is straight forward.

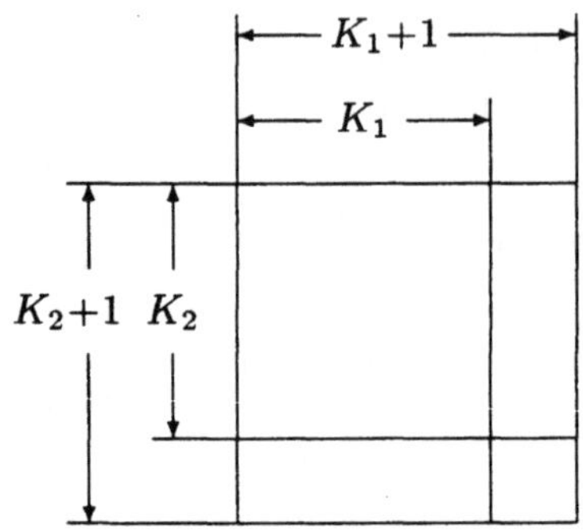

Figure 5.7 Two dimensional iterative ARQ protocol.

The feature of this ARQ protocol is that the probability of retransmission of the i-th stage is a monotonous decreasing function of i, but the probability of undetected errors can be made as small as desired by increasing M. In fact, it has been shown in [87] for the code and the ARQ protocol determined in such a way that the following upper bound of the final probability of undetected error is true.

$$P_f(M) \leq \frac{1}{2}x^{2^{M-1}} \text{ for } M \geq 2 \tag{5.79}$$

where $x = p/(1-p)$. At the same time we obtain the following lower bounds on the throughput efficiency [87]

$$\eta_M \geq 1 - 3.11\sqrt{x} - 1.6x \text{ for } M \geq 2 \tag{5.80}$$

for the optimally chosen block lengths n_i, $i = 1, 2, \ldots, M$ on each stage, which are

$$n_1 = \frac{1}{\sqrt{2.414x}}, n_2 = 3.414n_1, n_3 = \frac{n_2^2}{3.414}, n_m = 0.707n_{m-1}^2 \text{ for } m \geq 4. \tag{5.81}$$

It was shown in [87] that the single parity-check codes are superior to some other codes for obtaining the largest TE for a given $P_f(M)$.

Unfortunately, the iterative algorithm is not suitable if we have a noisy return channel, because we need to distingish the number of blocks at each stage.

Fink and Mukhametshina [61] proposed *recurrent stochastic coding* (RSC) to keep the strong property for real return channels. We will describe this approach very briefly. Each block of length n is divided into l subblocks of length

n_0, so that $n = ln_0$. One of these subblocks is a check subblock, and the remaining are information subblocks. The check subblock is formed by summation (modulo 2) of several information subblocks of the previous block. Each of these sums contains at most one subblock from every block, but each information subblock can enter into s check sums, where $s \geq 1$. In Figure 5.8 we show an example of RSC for $l = 5$ and $s = 1$. The dotted lines combines subblocks belonging to a common sum.

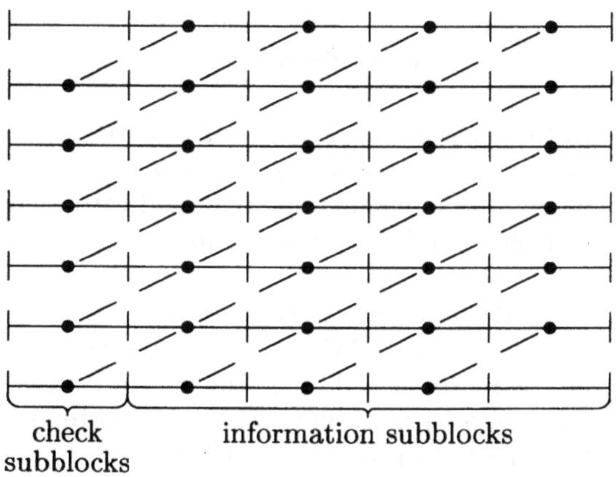

Figure 5.8 An example of recurrent stocastic coding with $l = 5$ and $s = 1$.

In the general case, the number of correlated code blocks will be $L = s(l-1)+1$. After coding in this manner each code block is executed by two stochastic transforms (SATC+SMTC, see 4.2) at the transmitting side and by inverse stochastic transforms at the receiving side that provides the "uniform" distribution (1.44) of the probabilities of the error patterns for any channel.

Thus each block is checked for the "engagement" with L previous blocks. If this checking is not true, then we discard this block and send an NAK signal over the return channel. This procedure continues until we receive two coinciding retransmitted blocks with detected errors (without "engagement" with the previous L blocks). It is easy to show that after the stochastic transforms (and only after them) the more likely event is that this block has been received correctly but one of the previous blocks is erroneous. Therefore, at this point we begin to send NAK to the previous block and resend it in a similar manner, and so on up to the last block, which is stored in the buffer. The main result obtained in [61] is that for a RSC with the given protocol, the probability of undetected errors approaches zero with increasing s and simultaneously the throughput efficiency remains lower bounded by some value depending on the RSC parameters and $P_{\text{ef}}(n)$ for the given channel. So a "strong" property is true for this method which is superior to Jacobs' approach [87] because it is suitable also for noisy return channels. Moreover, RSC guarantees a given reliability for any symmetric channels with memory if we use jointly SATC+SMTC, and a TE which depends on the $P_{\text{ef}}(n)$ only (see Chapter 1). This method is superiour to other methods in the probabilities of deletion, which we can get much less than 2^{-n_0} where n_0 is the number of check symbols in the check subblock. RSC can be combined with the stop-and-wait or selective repeat protocols although this requires some modifications.

The main shortcoming of this method is the use of stochastic transforms which require more complex hardware (or software). It also requires a significant memory size at the receiving and transmitting sides, and rather complex logic to implement the protocol. For details, we refer to [61].

A special class of protocols allow the use of concatenated codes with some of the protocols described above (WG, GBM, SR, etc). We will use a common term for them, *concatenated protocols.* Let us consider a concatenated code C with an inner $[n_1, k_1, d_1]$ code C_1 and outer $[n_2, k_2; q]$ shortend RS code C_2. The inner code is used to correct at most t errors (where $2t < d_1$) and it further detects errors. The outer code corrects at most $S_0 = n_2 - k_2 - l$ erasures (of inner code blocks), where $0 \leq l \leq n_2 - k_2$, and also detects errors.

An upper bound on $P_{\text{ue}}(C_1)$ for this CC, on any LBC-n_1, is given in Theorem 4.2.6. (We assume for each i-th inner block that the channel is a BSC with symbol error probability $p_i \in [0, \frac{1}{2}]$). We can use this CC with two main ARQ schemes:

A) Request for retransmission of a full block of length $n = n_1 n_2$ if more than S_0 inner blocks are erased or the outer code detects errors.

B) Request for retransmission of the inner blocks until they are received without detected errors; request a full block of the CC if the outer code detects errors (we use $S_0 = 0$).

Scheme B has the following modification:
B′) Send NAK signals to the inner blocks at most α times. If the inner code still detects errors after α retransmissions, this block is erased. The outer code is used to correct at most S_0 erasures and detect errors as for scheme A.

It is easy to derive the following upper bound for the probability of a request for a full code block when scheme A is used on the LBC-n_1 without memory:

$$P_{\rm r}^{\rm A} \leq 1 - \sum_{i=0}^{n_2-k_2-l} \binom{n_2}{i} \left\{ \left(1 - P_{\rm c}^{(t)}(C_1) - P_{\rm ue}^{(t)}(C_1)\right)^i P_{\rm c}^{(t)}(C_1)^{n_2-i} \right\}, \quad (5.82)$$

where

- $P_{\rm c}^{(t)}(C_1) = P(n,t)$ is the average probability of having at most t errors for the inner code block,
- $P_{\rm ue}^{(t)}(C_1)$ is the probability of an undetected error for C_1 after correction of at most t errors.

We get an upper bound in (5.82) because we have ignored the possiblity that the outer code has not detected errors that may be present. The probability of this is very small.

For scheme B we have a trivial expression for the probability of a request for an inner code block:

$$P_{\rm rin}^{\rm B} = 1 - P_{\rm c}^{(t)}(C_1) - P_{\rm ue}^{(t)}(C_1). \quad (5.83)$$

The block of the outer code in scheme B will only be requested when we have receiced all the inner blocks without detected errors, but the outer code have detected an error. As before, we ignore the probability that the outer code has undetected errors. Since the probability of undetected error after retransmission is equal to the final probability, we get the following upper bound from (5.7).

$$P_{\rm rout}^{\rm B} \leq 1 - \left\{ \frac{P_{\rm c}^{(t)}(C_1)}{P_{\rm c}^{(t)}(C_1) + P_{\rm ue}^{(t)}(C_1)} \right\}^{n_2}. \quad (5.84)$$

For scheme B′ the probability of a request for an inner code block coincides with the one given by (5.83), and the probability of a request for the outer code block has the following upper bound:

$$P_{\rm rout}^{\rm B'} = 1 - \sum_{i=0}^{n_2-k_2-l} \binom{n_2}{i} \bar{P}^{(t)}(C_1,\alpha)^i P^{(t)}(C_1,\alpha)^{n_2-i}, \quad (5.85)$$

where

- $P^{(t)}(C_1, \alpha)$ is the probability to have at most t errors on the inner block code after at most α retransmissions,
- $\bar{P}^{(t)}(C_1, \alpha)$ is the probability of a detected error in an inner block after α retransmissions.

In turn, the probabilities $P^{(t)}(C_1, \alpha)$ and $\bar{P}^{(t)}(C_1, \alpha)$ can be calculated as follows.

$$P^{(t)}(C_1, \alpha) = P_c^{(t)}(C_1) \sum_{i=0}^{\alpha} \left(1 - P_c^{(t)}(C_1) - P_{ue}^{(t)}(C_1)\right)^i$$

$$= P_c^{(t)}(C_1)\left(1 - P_c^{(t)}(C_1) - P_{ue}^{(t)}(C_1)\right)^{\alpha+1} / \left(P_c^{(t)}(C_1) + P_{ue}^{(t)}(C_1)\right)$$

$$\approx 1 - \left(1 - P_c^{(t)}(C_1)\right)^{\alpha+1}, \tag{5.86}$$

$$\bar{P}^{(t)}(C_1, \alpha) = \left(1 - P_c^{(t)}(C_1) - P_{ue}^{(t)}(C_1)\right)^{\alpha+1}. \tag{5.87}$$

To calculate the reliability and throughput efficiency of the concatenated strategy, it is necessary to specify what kind of protocol will be used for the retransmissions of the inner and outer code blocks. While scheme A is compatible with any protocol (WG, GBM, SR), the schemes B and B′ require SR versions of the protocol, because they have two states of requests — the inner code and the outer code — it is necessary to distinguish NAK signals of these states.

We consider only the perfect selective repeat protocol with unlimited size of the buffer in combination with these concatenated strategies.

The reliability of scheme A can be found from (5.49) by changing P_{ue} to $P_{ue}(CC)$ and P_{ef} to $(1 - P_r^A)$. This gives

$$P_f(CC) \leq \frac{P_{ue}(CC)}{P_{ue}(CC) + (1 - P_r^A)} \tag{5.88}$$

where $P_{ue}(CC)$ can be taken from Theorem 4.2.6 or (4.29) for any LBC-s (in particular any BSC) if the parameter s is at least equal to n_1. Note that for the equivalent error probability (P_{eq}) calculation we have to put $k = k_1 k_2$ into (5.50).

The throughput effiency can be calculated by (5.52), where

$$\Delta \geq \left(1 - P_{\rm r}^{\rm A}\right)\left(1 - P_{\rm r}'^{\rm A}\right)$$

and " ′ " denotes the corresponding value in the return channel, so we get

$$\eta_{\rm A} \geq \frac{k_1 k_2}{n_1 n_2}\left(1 - P_{\rm r}^{\rm A}\right)\left(1 - P_{\rm r}'^{\rm A}\right). \tag{5.89}$$

It is obvious that the presence of several service symbols among the information symbols can be neglected if $k_1 k_2$ is sufficiently large.

For scheme B formula (5.88) for $P_{\rm f}(CC)$ will be true if we change $P_{\rm r}^{\rm A}$ to $P_{\rm rout}^{\rm B}$. Using the state graph for scheme B and the rules given in Appendix 5.6, it is easy to show that the generating function corresponding to a transmission concatenated code block will be

$$f(x) = \frac{\Delta_{\rm out}}{\left\{\frac{1-\Delta_{\rm in}x}{\Delta_{\rm in}x}\right\}^{n_2} - (1 - \Delta_{\rm out})}, \tag{5.90}$$

where $\Delta_{\rm in}$ and $\Delta_{\rm out}$ are the probabilities that a given transmission of the inner and outer code block respectively will be the last attempt to convey them.

Taking into account the relation (5.25), we get

$$\begin{aligned} \Delta_{\rm in} &\geq \left(1 - P_{\rm rin}^{\rm B}\right)\left(1 - P_{\rm rin}'^{\rm B}\right), \\ \Delta_{\rm out} &\geq \left(1 - P_{\rm rout}^{\rm B}\right)\left(1 - P_{\rm rout}'^{\rm B}\right). \end{aligned} \tag{5.91}$$

Using (5.103), (5.90), and (5.91) we get the following upper bound for the throughput efficiency of scheme B.

$$\eta_{\rm B} \geq \frac{k_1 k_2}{n_1 n_2}\left(1 - P_{\rm rin}^{\rm B}\right)\left(1 - P_{\rm rin}'^{\rm B}\right)\left(1 - P_{\rm rout}^{\rm B}\right)\left(1 - P_{\rm rout}'^{\rm B}\right) \tag{5.92}$$

where " ′ " denotes the corresponding probabilities in the return channel.

(To derive (5.92) we assume that the effect of presence of the service symbols (addresses of the requested blocks) among the information symbols of the inner code blocks can be neglected, but this assumption is not quite adequate for short block lengths of the inner code. Therefore it would be better to change k_1 to $k_1 - l$ where l is the number of address bits).

Note that we could avoid the graph approach here for the proof of (5.92) using the well known result that the expection of a sum of a random number of random variables is equal to the product of the item expectation by the expectation of the number of items.

Example. Let us consider a concatenated code with $n_1 = 64$, $k_1 = 36$, $t = 4$, $n_2 = 16$, $k_2 = 15$, and $l = 1$. For strategy B using the modified (5.88) and (5.92) for a noiseless return channel and forward BSC with $p = 5 \cdot 10^{-2}$, we get $P_{\text{eq}} \approx 7 \cdot 10^{-7}$ and $\eta_{\text{B}} \geq 0.394$. Whereas for an ordinary binary linear code with combined t-error correction and error detection after $[n, k]$ parameter optimization for $t = 4$ we get $n_{\text{opt}} = 72$, $k_{\text{opt}} = 32$, $P_{\text{eq}} = 5 \cdot 10^{-8}$, and $\eta \approx 0.315$. The TE here is only slightly less than that for the previous concatenated protocol, but its implementation will be significantly simpler. Therefore, a concatenated strategy can be considered superior to an ordinary linear code only from the point of view of a simpler calculation of the probabilities of undetected errors. In fact, it is not easy to find the weight distribution for a linear code having more than 40 check symbols, and this is necessary to calculate P_{ue} for the code, whereas the $P_{\text{ue}}(CC)$ for the concateneted code can easily be found from Theorem 4.2.6 or (4.29).

A significantly different situation may occur when multitone modems are used and there is a selective fading on every tone. (Such a situation is typical for mobile satellite communication or for high frequency channels with long communication distances). We get the LBC model (see Chapter 1) if we place the symbols from one inner code block on different tones. Then we can use the bounds in Theorem 4.2.6 or (4.29) to estimate $P_{\text{ue}}(C)$, but we have no tool to solve the similar problem for ordinary linear codes. It may appear that there is a fool-proof solution to this problem — to place all the symbols of any ordinary linear code on the same tone because this results in the BSC model. Unfortunately, this choice requires a very large size of the receiving buffer to convey the code blocks to the user in the correct order, namely the first block is placed on the first tone, at the same time the second block is placed on the second tone and so on. To prove this, let us consider a situation where we have n_{f} tones, a code of block length n_{t}, and a round trip delay of n_{rt} code blocks. We also assume that the forward channel has independent errors between tones and the return channel is noiseless. To provide a correct order of the blocks it is necessary to store n_{rt} blocks at each of the n_{f} tones if there are detected errors on at least one tone. This results in the following relation for the average size of the receiving buffer (in number of symbols)

$$V_{\text{av}} = A \sum_{s=1}^{\infty} sP(A_s),$$

where $A = n_\mathrm{t} n_\mathrm{f} n_\mathrm{rt}$ and $P(A_s)$ is the probability that at least one code block out of n_f tones will be resent exactly s times and the last block will be transmitted less than s times. This can be obtained as follows:

$$\begin{aligned} P(A_s) &= \sum_{i=1}^{n_\mathrm{f}} \binom{n_\mathrm{f}}{i} \left((1 - P_\mathrm{d})\, P_\mathrm{d}^s \right)^i \left(\sum_{j=0}^{s-1} (1 - P_\mathrm{d})\, P_\mathrm{d}^j \right)^{n_\mathrm{f} - i} \\ &= \left(1 - P_\mathrm{d}^{s+1}\right)^{n_\mathrm{f}} - \left(1 - P_\mathrm{d}^s\right)^{n_\mathrm{f}} . \end{aligned}$$

Putting this expression into the formula for V_av we get

$$V_\mathrm{av} = A \sum_{s=1}^{\infty} s \left\{ \left(1 - P_\mathrm{d}^{s+1}\right)^{n_\mathrm{f}} - \left(1 - P_\mathrm{d}^s\right)^{n_\mathrm{f}} \right\} . \tag{5.93}$$

If we use the above mentioned CC with strategy A having parameters ($n_1 = n_t$, k_1, t, $n_2 = n_\mathrm{f}$, k_2, l) we get

$$V_\mathrm{av}^\mathrm{A} = n_\mathrm{t} n_\mathrm{f} \left(1 - P_\mathrm{r}^\mathrm{A}\right) \sum_{i=1}^{\infty} n_\mathrm{rt} i \left(P_\mathrm{r}^\mathrm{A}\right)^i = n_\mathrm{t} n_\mathrm{f} n_\mathrm{rt} P_\mathrm{r}^\mathrm{A} / \left(1 - P_\mathrm{r}^\mathrm{A}\right) . \tag{5.94}$$

For the average size of the buffer for strategy B we have

$$V_\mathrm{av}^\mathrm{B} = V_\mathrm{av} / \left(1 - P_\mathrm{r}^\mathrm{B}\right) , \tag{5.95}$$

where V_av is determined by (5.93).

The expression for the average size of the buffer for strategy B' can easily be derived as

$$V_\mathrm{av}^{\mathrm{B}'} = \frac{A}{\left(1 - P_\mathrm{rout}^{\mathrm{B}'}\right)} \left\{ \alpha + \sum_{s=1}^{\alpha-1} (s - \alpha) \left[\left(1 - P_\mathrm{d}^{s+1}\right)^{n_\mathrm{f}} - \left(1 - P_\mathrm{d}^s\right)^{n_\mathrm{f}} \right] \right\} . \tag{5.96}$$

To illustrate the advantages and disadvantages of these cases, we consider an example.

Example. Consider a channel model where on each tone we have CRFC (see Chapter 1), and the parameters describing this model are $h^2 = 98$ and $\alpha = 1/2$. We take the following parameters for the ordinary linear code: $n_\mathrm{t} = 60$, $k = 40$, and $n_\mathrm{rt} = 18$. Then we get $P_\mathrm{eq} \leq 2.5 \cdot 10^{-8}$ (if the code is a good code), $\eta \approx 0.61$, and $V_\mathrm{av} \approx 12$ kbits.

For the concatenated strategy A we take a code with $n_1 = 60$, $k_1 = 47$, $t = 0$ (it is useless to correct errors in the CRFC mode), $l = 2$, $n_2 = 64$, $k_2 = 54$.

Then we obtain $P_{eq} \leq 5 \cdot 10^{-11}$, $\eta = 0.6392$, $V_{av}^{A} \approx 284$ bits. We can see a significant difference in buffer sizes for an ordinary code and for CC. The latter approach is superior in TE also. If we take concatenated strategy B with parameters $n_1 = 60$, $k_1 = 47$, $t = 0$, $n_2 = 64$, $k_2 = 60$, $l = 1$ we obtain $P_{eq} \approx 7 \cdot 10^{-11}$, $\eta = 0.705$, and $V_{av}^{B'} = 12$ kbits. We see that strategy B is superior to strategy A for TE but for average buffer size the situation is opposite. Fuller conclusions can be given after a joint optimization of code parameters, strategy, and protocol (see Chapter 6).

The "hard" startegy A can be changed to a more "soft" one as follows (see Zyablov [213]). Let us take some concatenated code C with inner $[n_1, k_1]$ code C_1, and outer Reed-Solomon $[n_2, k_2]$ code C_2. First we transmit k_2 information q-ary symbols of the outer code (which are blocks of the inner code). Then we begin transmission of the outer code check symbols one by one. At the receiving end errors are corrected and detected in the inner code blocks and the erasures are corrected by the outer code. If the number of erased $q = 2^{k-1}$-ary symbols of the outer code is not more than the $[n_2', k_2]$ code can correct after the transmission of $r_2' - k_2$ check symbols, and the errors are not detected by the outer code, this (n_2', k_2) code block is accepted and the ACN signal is transmitted over the return channel to avoid further transmissions of q-ary check symbols. Otherwise an NAK signal is transmitted over the return channel to continue symbol transmission up to $n_2 = n$. Then we discard this $[n_2, k_2]$ code block and begin to transmit again k_2 information q-ary symbols, and so on. This protocol is good also for channels with independent bursts of errors which are located in inner code blocks. It is easy to derive an expression for TE with this protocol. However, there are significant difficulties in this application with a noisy return channel and this makes it less attractive. In fact, if a ACN signal has been transmitted, but there were detected errors in the return channel, then the transmission of check symbols will continue, but at the receiving end one may believe that these are information symbols of a new concatenated code block. This situation can be remedied by providing each inner code block with a number (modulo n_2), but this obviously decreases the TE.

5.4 PROTOCOLS WITH INFORMATION FEEDBACK

Until now we have only considered protocols with decision feedback which are systems making the decision about a transmission of a new code block or retransmission of an old block at the *receiving* end. In this section we consider systems where this decision is made at the *transmitting* end. There are many kinds of information feedback protocols (see Kanevsky [9]), but the most important of them which provide both high reliability of the information transmission and good throughput efficiency is the following one. Let us take some systematic linear $[n, k, d]$ code C. The information emitted from the source is coded by the code C, but over the forward channel only the k information symbols are transmitted. At the receiving end the corresponding $r = n - k$ check symbols are determined from the received information symbols (possibly corrupted by noise), and these check symbols are transmitted over the return channel. When they are received they are compared with the original check symbols. If they all coincide, the next information block is transmitted over the forward channel with an additional ACN signal; otherwise we repeat the previous information block with an additional NAK signal. Let $e^{(1)} = (e_1^{(1)}, e_2^{(1)}, \ldots, e_k^{(1)})$ denote the error pattern that occured in the forward channel during transmission of the information symbols, and let $e^{(2)} = (e_1^{(2)}, e_2^{(2)}, \ldots, e_r^{(2)})$ the error pattern that occured in the return channel during transmission of the check symbols. Then errors are not detected if and only if

$$b_j = \tilde{b}_j \quad \text{for } j = 1, 2, \ldots, r. \tag{5.97}$$

Here

$$b_j = \sum_{i=1}^{k} \beta_{ij} a_i, \tag{5.98}$$

where $(a_1, a_2, \ldots, a_k)$ are the information symbols and β_{ij} are the elements of the parity check matrix,

$$b'_j = \sum_{i=1}^{k} \beta_{ij} \left(a_i + e_i^{(1)} \right), \tag{5.99}$$

$$\tilde{b}_j = b'_j + e_j^{(2)}. \tag{5.100}$$

(As usual summation is modulo 2). Combining (5.97)-(5.100) we get

$$e_j^{(2)} = \sum_{i=1}^{k} \beta_{ij} e_i^{(1)} \quad \text{for } j = 1, 2, \ldots, r. \tag{5.101}$$

It follows from (5.101) that errors are not detected for this protocol if and only it the error pattern $(e^{(1)}|e^{(2)})$ is a code word in C. Thus, if both the forward and the return channels are BSC_p and they are independent, then we obtain the same probability of undeteced error for this protocol as we would get using the same code C in a decision feedback protocol (see Chapters 1-3). But it is another question if these channels are different, for example if both are BSC, but with distinct symbols error probabilities p_f and p_r; this gives a special LBC (see Chapter 1 and 4.2). In particular, if $p_r = 0$, i.e. we have a noiseless return channel, the probability of undeteced error is given by (4.21). In the general case it is not easy to find an expression for P_{ue}.

It is obvious that for duplex systems and a noiseless return channel we will have the same throughput efficiency for all basic types of protocols (WG, GBM, SR) as for decision feedback systems. But the situation changes drastically if ACN and NAK signals can be erroneous.

One feature of this protocol is the impossibility to detect errors at the receiver side. Therefore if a NAK signal is transformed to an ACN signal, the erroneous information block will not be erased but passed on to the user. Since the sender of an NAK can recognize (with probability at most P_{ue} that this signal was corrupted and that the erroneous block was not erased, it would try to correct this situation. There are several strategies which allow us to "smooth" drastically the consequences of errors in the NAK or ACN signals (see Chang [47]). But most of them have the property that any block stored at the receiver side can be erased in the future with some non-zero probability. This results in the necessity to have a very large memory at the receiving side even for a zero round-trip delay.

This effect is most pronounced at the channels with varying parameters when there are time intervals with bad channel states, with symbol error probabilities close to 1/2 (i.e. near to the cut of the channel). We will not describe these protocols in any detail, because as far as we know they have not found practical implementations, but it does not preclude the possibility that some modification of the information feedback strategy may appear in future applications.

5.5 CONCLUSIONS

In this chapter we have considered a number of different protocols for feedback systems. As a rule we have been interested in the following problems:

- to describe an algorithm at the receiving and the transmitting sides to provide this protocol,
- to give formulas for the reliability of information after using this protocol (taking into account the probability of information shifts (deletions and insertions)),

- to present formulas to calculate the throughput efficiency (TE) for the given protocol,
- to estimate the buffer sizes at the receiving and transmitting sides for a given round-trip-delay in the channel.

Except for some complex algorithms, we have assumed that the return channel may be noisy.

Calculation of the TE has been done by a general method based on signal flow graph techniques. The method is described in Appendix 5.6.

This chapter does not claim to be a handbook on protocols. There are several reasons for this. First of all, there are presently more than twenty different protocols, and to describe them all in sufficient detail necessary for an implementation we would need to increase the size of this chapter at least by a factor three and make it much harder to read. Secondly, some of these modifications can be derived by the reader without difficulties. Finally, some of the protocols contain, to our knowledge, unsolved problems.

To simplify practical application of these results we have summarized the main formulas for TE in Table 5.3.

The reliability of information transmission with ARQ protocols can be estimated by formulas (5.7), (5.11), (5.12), (5.18), and (5.19) except for the iterative algorithm (5.79) and concatenated protocols (5.88). We can also roughly approximate the probabilities of information shift (deletion and insertion) by P_{ue}.

In the text of this chapter we have made some comparisons of the TE of different protocols for a given reliability. The conclusions depend very much on the parameters of the channels for information transmission and restrictions on the complexity of the hardware or software implementations of the given protocol. More detailed comparisons will be made in the following chapter where we try to get the best value of TE over different protocols for a given reliability.

Remark. Throughout this chapter we have been concerned with descriptions of protocols dedicated to a *point-to-point communication.* Another theme is *communication through data networks.* In this case we have different levels of error control — physical, link, network, transport etc. There are protocols to carry out this function for these levels. for example, for the link level we

Table 5.3 A list of the main formulas for TE calculations for different ARQ protocols.

Protocol	Formula number			
	BSC, noiseless return channel	BSC, noisy return channel	Partial error correction	Channel with memory
Stop-and-wait	(5.29)	(5.29)	(5.30)	
Go-back-M	(5.37) (5.39)	(5.37) (5.38)	(5.47) (5.48)	(5.46)
Selective repeat	(5.52) (5.53)	(5.52) (5.53)	(5.47) (5.48)	
Selective repeat + go-back-M	(5.56)		(5.47) (5.48)	
Address confirmation	(5.57)	(5.57) (5.58)	(5.47) (5.48)	
Selective repeat for session transmisson	(5.67)	(5.68)	(5.47) (5.48)	
Type II hybrid ARQ	(5.76) (5.77)		(5.76) (5.77)	
Iterative algorithm	(5.80)			
Concatenated protocols (A,B,B′)	(5.89) (5.92)	(5.89) (5.92)	(5.47) (5.48)	(5.89) (5.92)

have the HDLC protocol which is part of CCITT X.25.3. In Bertsekas and Gallager's book [2] the readers can find ARQ protocols which are very similar to some of those described above and with the same names: stop-and-wait (in particular Arpanet ARQ), go-back-n, selective repeat, etc. The main difference to the situation described in the point-to-point protocols described above is in the following conditions typical for data networks:

- existence of lower and upper bounds for the length of code blocks, i.e. dimension of frames or packets (as a rule they are large, the standard length of a frame is 128 kbits),
- the main criterion for time efficiency is not the average information rate but the average time to deliver a packet (or the distribution function of

this time) for a given reliability (the probability to receive a packet without errors).

The first condition results in the possibility to insert frame numbers into the blocks, because this redundancy can be neglected in view of the long block length. This circumstance allows us to use simpler methods to remove deletions and insertions of frames for noisy return channels. So in [2] a stop-and-wait protocol is given for this purpose. It is shown that numbering modulo 2 is sufficient to avoid deletions and insertions if the undetected error probability can be made neglectably small.

To obtain this condition, the numbers of the following waiting packets should be sent over the return channel. Similar numbering modulo n can be used also for a go-back-n protocol. If the length of the code blocks is small, the additional address symbols decrease code rate and therefore in this chapter we have suggested other methods to avoid deletions and insertions. But for both point-to-point protocols and data network protocols the following conclution is true: for an appropriately chosen protocol the probabilities of deletions and insertions can be made the same order as P_{ue} for a given code. In the next chapter we will optimize codes and protocols, mainly for point-to-point communication, but some of these methods will be valid also for data communication if we take into account the above mentioned conditions. The fact that the attention is focused on point-to-point communication can be explained as follows. Firstly, to get from some terminal to a transport network can often be possible only via some noisy channel (high frequency, satellite, and others) without X.25 application, and then we can use point-to-point ARQ protocols. Secondly, there may be special leased channels for mobile or personal communication. We believe that these are sufficient reasons to restrict our consideration to this case.

5.6 APPENDIX. SIGNAL FLOW GRAPHS AND THEIR APPLICATION TO PROTOCOL ANALYSIS

Consider a directed graph (A, W) given by a set A of nodes and a set of W of directed branches. Every node of the graph corresponds to some state of a feedback system protocol, a branch between two nodes means that it is possible to get from one to the other in the given direction. For each branch there is assosiated a probability that this branch is choosen. Further, there is assosiated a non-negative integer random variable (this is the number of "steps" which are necessary to do to go from one state to another). This random variable ξ

can be described by a generating function (a power series):

$$f_\xi(x) = f(x) = \sum_{k=0}^{\infty} P_k x^k, \tag{5.102}$$

where $P_k = Pr(\xi = k)$. It is well known (see e.g. Huggins [84]) that this function has the following properties:

$$\begin{array}{l} f(1) = \sum_{k=0}^{\infty} P_k = 1, \\ E(\xi) = f'(1), \\ Var(\xi) = f''(1) + f'(1) - \left(f'(1)\right)^2. \end{array} \tag{5.103}$$

We remark that if we introduce the function

$$q(x) = \frac{1 - f(x)}{1 - x}$$

we get the following expressions which are sometimes easier to calculate (see Feller [6] page 249).

$$\begin{array}{l} E(\xi) = q(1), \\ Var(\xi) = 2q'(1) + q(1) - \left(q(1)\right)^2. \end{array} \tag{5.104}$$

The function $q(x) = \sum_{k=0}^{\infty} q_k x^k$ is the generating function for the tails of ξ:

$$q_k = p_{k+1} + p_{k+2} + \cdots.$$

We have $f_{\xi+\eta}(x) = f_\xi(x) f_\eta(x)$ if and only if ξ and η are independent random variables.

Consider a pair (ξ, η) of random variables. We define the two-dimensional generating function

$$\phi_{\xi,\eta}(x, y) = \phi(x, y) = \sum_{k=0}^{\infty} \sum_{l=0}^{\infty} P_{kl} x^k y^l, \tag{5.105}$$

where $P_{kl} = Pr(\xi = k, \eta = l)$. It can be shown that for any pair (ξ, η) of random variables we have

$$f_{\xi+\eta}(x) = \phi_{\xi,\eta}(x, x). \tag{5.106}$$

Let us fix two states corresponding to the nodes i and j of the signal graph. If we know the generating function $f(x)$ for the number of steps to go from i to j, we can solve the following problems (using (5.103)):

- to find the probability to get from i to j in k steps, this is the coefficient of x^k in $f(x)$,
- to calculate the expected number of steps to go from i to j,
- to calculate the variance of the number of steps to go from i to j.

These are very important characteristics for a description of feedback system protocols and we will return to them a little bit later, but first we will study a method for calculation of these generating functions between any two nodes. The key point is to give some rules for graph transformations which allow us to reduce any complex graph between two nodes to a simpler graph with the same generating function. The basic reduction rules, taken from [84], are given in Figure 5.9.

The transformations of the original graph with the use of these rules are illustrated in Figure 5.10, which shows the steps in the reduction of a graph into a single branch.

The final generating function

$$f(x) = \frac{f_1(x)\left\{f_5(x) + \frac{f_2(x)f_4(x)}{1-f_3(x)}\right\}}{1 - f_6(x)\left\{f_5(x) + \frac{f_2(x)f_4(x)}{1-f_3(x)}\right\}} \qquad (5.107)$$

is the generating function of the original graph.

We can use a generating function approach to find P_{f}, P_{in}, P_{de}, or the throughput efficiency for different protocols of the ARQ. Depending on the problem we want to solve, we assign different generating functions on the branches. As an example of such an approch, consider the problem of calculating P_{de} for the stop-and-wait protocol with a noisy return channel. The graph corresponding to this protocol is shown in Figure 5.11.

Here we have the following set of nodes corresponding to the states of the ARQ system:

- 1 — transmission of the first code information block,
- 1′ — receiving the first block without detected errors,
- 1″ — receiving the first block with detected errors,

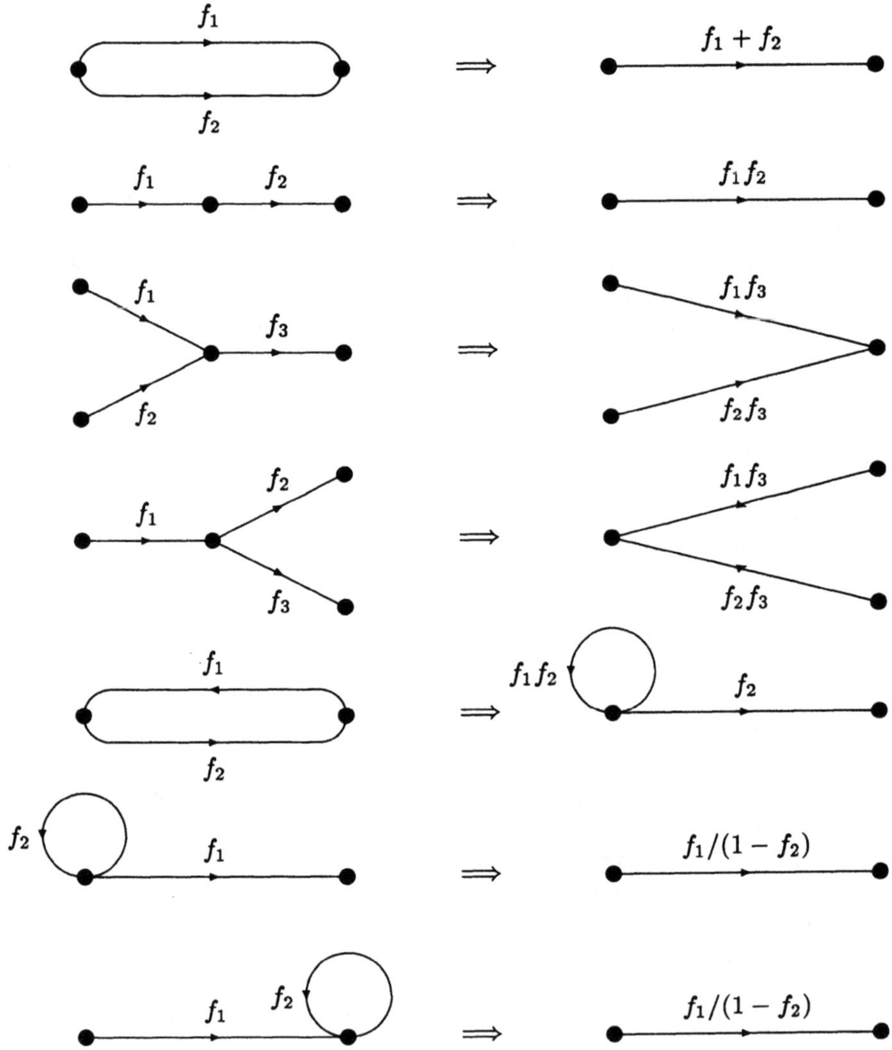

Figure 5.9 Graph reductions

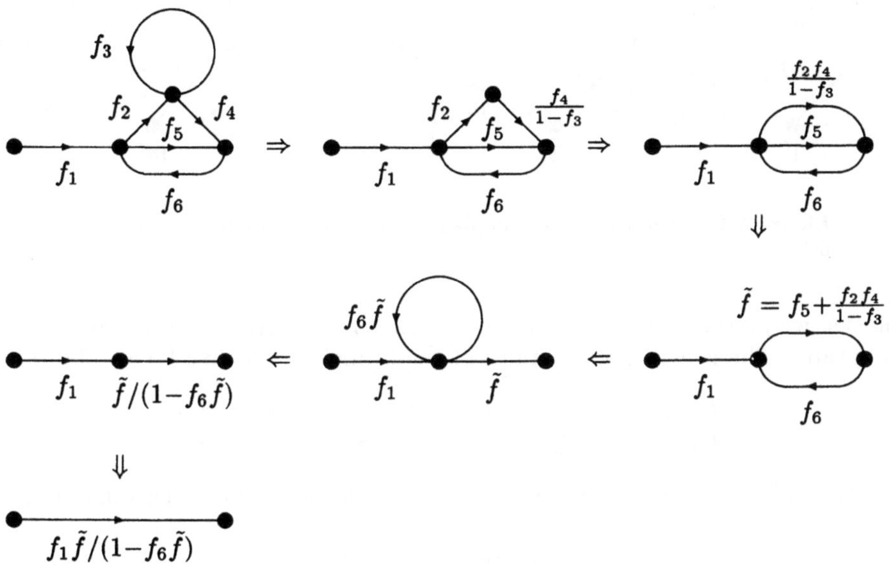

Figure 5.10 Example of the reduction of a graph.

- $2, 2', 2'', \ldots m, m', m''$ similarly for the second, third, etc. block,
- $m + 1$ — compeleted transmission of all m information blocks.

The labels of the branches show the probability of choosing this branch. They are:

- P_{d}, the probability of detected error,
- S_1, the probability of receiving ACN as a NAK in the return channel,
- S_2, the probability of receiving NAK as a ACN in the return channel.

It is clear that it is sufficient to find the generating function $f(x)$ for the first subgraph, the generating function of the full graph is then $f(x)^m$. In Figure 5.12 we derive $f(x)$.

Since we are only interested in calculating how many deletions of information blocks that will take place during the transmission to the user of the information

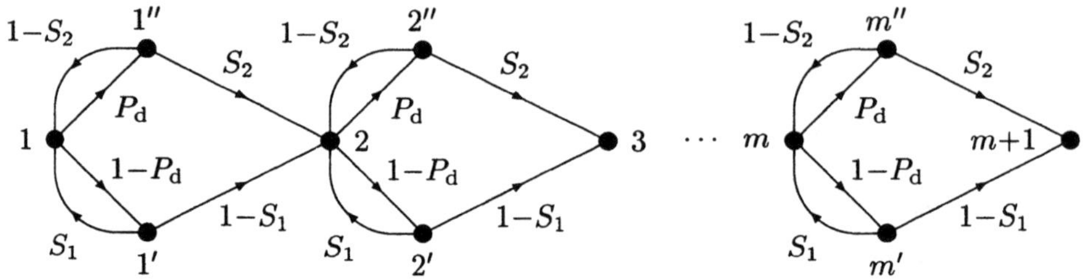

Figure 5.11 The graph of the stop-and-wait protocol with noisy return channel.

blocks, we prescribe 1 step to branch $(1'', 2)$ and 0 steps otherwise in the first subgraph. The generating functions of the branches will then be as follows:

$$\begin{aligned} f_1(x) &= 1 - S_2, \quad f_2(x) = P_\text{d}, \qquad f_3(x) = S_2 x, \\ f_4(x) &= 1 - P_\text{d}, \quad f_5(x) = 1 - S_1, \quad f_6(x) = S_1. \end{aligned}$$

Putting this in the expression for the graph generationg function given in Figure 5.12 we get

$$f(x) = \frac{P_\text{d} S_2 x + (1 - P_\text{d})(1 - S_1)}{1 - ((1 - S_2)P_\text{d} + (1 - P_\text{d})S_1)}. \tag{5.108}$$

Using (5.103) we get the expectation of deletion for transmission of m information blocks:

$$E(\xi) = \frac{d}{dx} (f(x))^m \Big|_{x=1} = \frac{m P_\text{d} S_2}{1 - ((1 - S_2)P_\text{d} + (1 - P_\text{d})S_1)}. \tag{5.109}$$

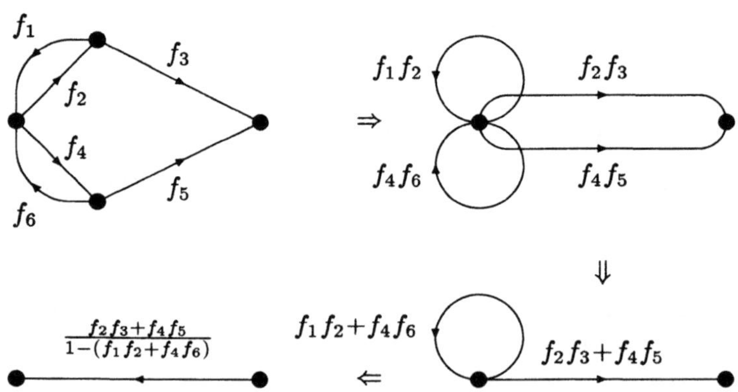

Figure 5.12 The partial graph and its transformation to the trivial graph.

Thus we can estimate the probability of deletion:

$$P_{\mathrm{de}} = \frac{P_{\mathrm{d}} S_2}{1 - ((1 - S_2)P_{\mathrm{d}} + (1 - P_{\mathrm{d}})S_1)}. \tag{5.110}$$

5.7 COMMENTS AND REFERENCES

5.2. The main criteria of FS efficiency are a combination of those given by Benice and Frey [35] and by Korzhik and Fink [10]. They are in line with the notations used in Lin and Costello [11] and in Odoul and Morgera [154].

5.3. The first three types of protocols were taken from Lin and Costello [11] and generalized here to two-way channels and channels with memory described by the FPMC model.

The formula (5.56) was quoted by Lin and Costello. The proof given here, using signal flow graphs, is new.

ARQ protocols with address confirmation is taken from Amosov and Aleksandrov [26]. The ARQ selective repeat protocol for session transmission systems has not been published before.

SR type II protocols are from Lin and Costello [11], Krishna and Morgera [123], and Lin and Gun [139]. Some of the formulas have been given new proofs using signal graphs. A significant contribution to hybrid ARQ typr-II protocols have been given by Wicker and Bartz [202], [203] where they use MDS and RM codes to improve TE, and by Rasmussen and Wicker [164]-[166] where trellis coded hybrid ARQ protocols are used.

Performance of the three basic ARQ protocils: SR, GBN, and SW on a channel with bursts was given by Lu and Chang [141].

The iterative protocol with the strong property ($P_{\mathrm{f}} \to 0$ when $n \to \infty$) is due to Jacobs [87]. The recurrent stochastic coding with the strong property has been developed by Fink and Mukhametshina [61].

Protocols with concatenated codes were investigated by Fujiwara and Kasami [65] and Korzhik, Dzubanov, Lesman, and Malev [117], [119]. A generalization of these protocols for noisy return channel and evaluation of TE for the Markov chain model was given by Odoul and Morgera [154].

Benelli [31] considered two new ARQ schemes, which results in a combination of the GBN and SR protocols, where the TE is evaluated through some approximative analytic method. He also gave a method for soft decision, using memory at the receiving side. The same author [32] described a new ARQ protocol based on concatenated coding.

A new selective ARQ protocol with finite length buffer is also described by Benelli [33]. The protocol achives results close to the SR protocol with infinite length buffers for high error rate in the communication channel.

The use of on-board processing on satellites has been studied by Hayashida [75]. The tandem type GBN scheme is an example of this approach. He showed that the round trip delay may be reduced by using this protocol effectively.

In this book we only consider point-to-point protocols, which have been studed extensively. In contrast, limited amount of work has been done on point-to-multipoint ARQ systems. This interesting field has been developed by e.g. Deng [54]. In that paper there is also a good bibliography of point-to-multipoint protocols.

Reliability information provided by sets of orthogonal check sums in a majority logic decoder for block codes is used in the type-I ARQ error control scheme by Rice and Wicker [170]. They show that the reliability performence of RM and other majority logic decodable codes can be improved at the expense of a small reduction in throughput.

Kim and Un [98] gave simple analytical formulas for the throughput efficency for the GBN and SR ARQ schemes, assuming that the forward and return channels have Markov error characteristics.

An efficient stop-and-wait ARQ protocol using soft error detection was developed by Fantacci [59].

Benelli [30] introduced a method to integerate the modulation in an ARQ scheme.

5.4. Protocols with information feedback are considered in more detail in Kanevsky [9] and Chang [47].

5.6. The description of signal graphs is from Huggins [84]. The application of graph theory to FS analysis has been developed by Chang [48] and others.

6

ERROR DETECTING CODE OPTIMIZATION FOR FEEDBACK SYSTEMS

6.1 BASIC CRITERIA FOR OPTIMIZATION

Let us consider the following formulation of the main problem considered in this chapter. We assume given some discrete channel (in the general case non-symmetric and with memory), a required reliability (e.g. bound on P_{eq} or P_{av}), and some ARQ protocol. We define the ***optimal error detecting code for the feedback system*** as the code which provides the highest throughput efficiency (TE) under the given conditions. As has been explained in Chapter 3, not all codes are proper for the BSC. A similar situation may occur for other channels, with a "less noisy" channel giving a higher P_{ue}. This is an undesirable situation. Therefore we introduce another condition: the code should provide the given reliability for any better states of the channel (for example for the BSC_p, the provided reliability should be provided for any $BSC_{p'}$ where $p' \leq p$.

A well known criterion for error *correcting* code efficiency is the *coding gain* (CG) given by

$$\theta_{(dB)} = 10 \log_{10} \frac{h_{wc}^2}{h_c^2} R, \tag{6.1}$$

where

- h_{wc} is the signal-to-noise ratio which provides the required reliability without coding,
- h_c is the signal-to-noise ratio which provides the required reliability using a code C,
- $R = k/n$ is code rate of the code C.

From (6.1) we see that the information transmission rate is the same whether we use the code C or not. Thus the code providing the highest coding gain can be called the *optimal error correcting code* for the given channel and required information transmission reliability.

In a similar manner one can define a CG criterion for error *detecting* codes in a feedback system. We only change the code rate R in (6.1) to η, the throughput efficiency. However, we will use this criterion only rarely. The reason for this is that as a rule a modem is given before error detecting code optimization and therefore the problem is to provide the required reliability at the cost of decreasing the data rate.

Many relations have been derived in Chapters 3 and 4 to calculate or estimate P_{ue} for some linear codes on the BSC and other channels. There we also introduced the notion of an optimal $[n, k]$ code which provides the minimum P_{ue} among all $[n, k]$ codes for a given channel. Unfortunately, these results do not allow us to find the strictly optimal error detecting code for feedback systems, because for most parameters n, k, there is no efficient algorithm to find an optimal code or its P_{ue}. Therefore, we have to look for good approximations. We call C an ϵ-optimal error detecting code for a feedback system (or just ϵ-optimal code) if its TE is at least a factor ϵ of the TE of the strong optimal code. If ϵ is close to 1 the code is *practically optimal.* This approach does not require the knowledge of P_{ue} for the strong optimal code, it is sufficient to have a good lower bound on P_{ue}.

As shown in Chapter 5, the value of TE depends very much on the type of protocol used. Therefore the choice of ϵ-optimal code will also depend on the type of protocol. We will give a general approach to finding ϵ-optimal codes for channels without memory in Section 6.2 and for channels with memory in Section 6.3. Partial error correction will also be considered. Several examples for the most important protocols will be given to compare their efficiencies and obtain the highest TE for a given reliability.

6.2 CODE OPTIMIZATION FOR CHANNELS WITHOUT MEMORY

We consider first pure error detecting codes on the BSC. Suppose that it is known how to find the optimal code for any p, n, and k (the $[n, k]$ code C with minimum $P_{\text{ue}}(C, p)$). Then, for any ARQ protocol the problem of code optimization is reduced to finding the parameters (n, k) such that the optimal $[n, k]$ code provides the highest TE for the given protocol under the condition that it satisfies the required reliability. The search can be done one k at a time; for a given k we find the least value of n which provides the required reliability for the optimal code. Thus we can find the optimal n and k which provides the highest TE. Next we have to check whether the chosen optimal code C provides the required reliability for all $p' < p$. By Theorem 3.1.4, this is the case if $p \leq d/n$ where d is the minimum distance of C.

However, we have no efficient algorithm to determine the optimal codes for all parameters (n, k). In practice we will therefore use some upper bound on $P_{\rm ue}[n, k, p]$, either one of the explicit bounds given in Chapter 3 or a bound found by searching through some class of $[n, k]$ codes for which we can calculate $P_{\rm ue}(C, p)$. To show the ϵ-optimality of the chosen code we can use the lower bounds on $P_{\rm ue}[n, k, p]$. The following theorem can be useful for estimating the ϵ-optimality of some codes.

Theorem 6.2.1 *If for a given ARQ protocol the TE can be expressed by a formula*

$$\eta = \frac{k}{n} f(n, p), \tag{6.2}$$

where $f(n, p) \leq 1$ and $f(n, p)$ is a monotonically decreasing function of n for any p, and a required reliability is given, then the TE for the optimal code of length n satisfies

$$\eta_{\rm opt} \leq \frac{k'}{n'} f(n, p), \tag{6.3}$$

if $n' > n$ and there is an $[n', k']$ Hamming code C' whose $P_{\rm ue}(C', p)$ does not satisfy the required reliability.

Proof: Suppose that there exists an $[n', n]$ Hamming code with $n' > n$ which does not satisfy the given reliability. Then no code with rate $R \geq k'/n'$ and length $n < n'$ can satisfy the required reliability since the Hanmming code is optimal and the error detecting capability decreases when the code length increases for a given code rate. This proves the theorem.

To speed up the optimization procedure, we can bound in various ways the range of values of n and k considered. We can start with some $k > 1$. We do not have to examine pairs (n, k) for which k/n is less than the TE of some previously considered code providing the given reliability. If n' is the minimum block length that provides the given reliability for some k', then we only have to consider block lengths above n' for $k'+1$. We can increase n or k by more than one in each step. The range of k may be bounded by some k_0. In particular, we can stop the search if we reach some k_0 such that in the previous step we found a TE η_0 such that $f(k_0,p) \leq \eta_0$, where f is the function in (6.2). (We note that these restrictions may result in us missing the optimal solution).

Note that initially in the code parameter optimization process we can use the simple upper bounds in Theorems 3.5.8, 3.5.9, or 3.5.10, where the minimum distance d in turn can be estimated e.g. by the Varshamov-Gilbert bound. If the resulting parameters (n,k) do not give the required reliability, the next step will be to consider some class of codes, e.g. CRC codes, and calculate P_{ue} from the weight distribution. One can either calculate and store the data for a (large) number of such codes beforehand and search the stored data during the optimization, or generate the codes as part of the optimization process. Finally, we estimate the highest ϵ for which the chosen code is ϵ-optimal (it will be at least η-optimal, where η is the TE of the code).

As a rule we have three typical situations for a BSC:

1. channel with low noise ($p \leq 10^{-5}$),
2. moderately noisy channel ($10^{-5} \leq p \leq 10^{-3}$),
3. very noisy channel ($p \geq 10^{-3}$).

The typical required reliability for data transmission is $P_r \approx 10^{-6} - 10^{-9}$. Therefore, the first candidate in case 1) is the single parity check code and in case 2) some Hamming code. For case 3) it is necessary to choose a code with higher redundancy, for example a shortened Hamming code. (Note that the required reliability, P_r is given as P_{eq} or P_{av}, but not as P_{ue}). In Figure 6.1 we give a block diagram for the optimization procedure for the required reliability given by P_{eq}. An implementation will use different bounds on P_{ue} as well as known weight distributions for special classes of codes as explained above. The procedure is valid for different ARQ protocols and different requirements for the reliability of the information transmission. We illustrate the procedure by several examples.

Example 1. We assume that the channel is a low noise BSC with $p = 10^{-5}$. The requirement is $P_{eq} \leq 10^{-9}$, and we use a perfect SR protocol with a noiseless return channel. Then the throughput efficiency is determined by formula (5.52) where $\Delta \approx (1-p)^n$.

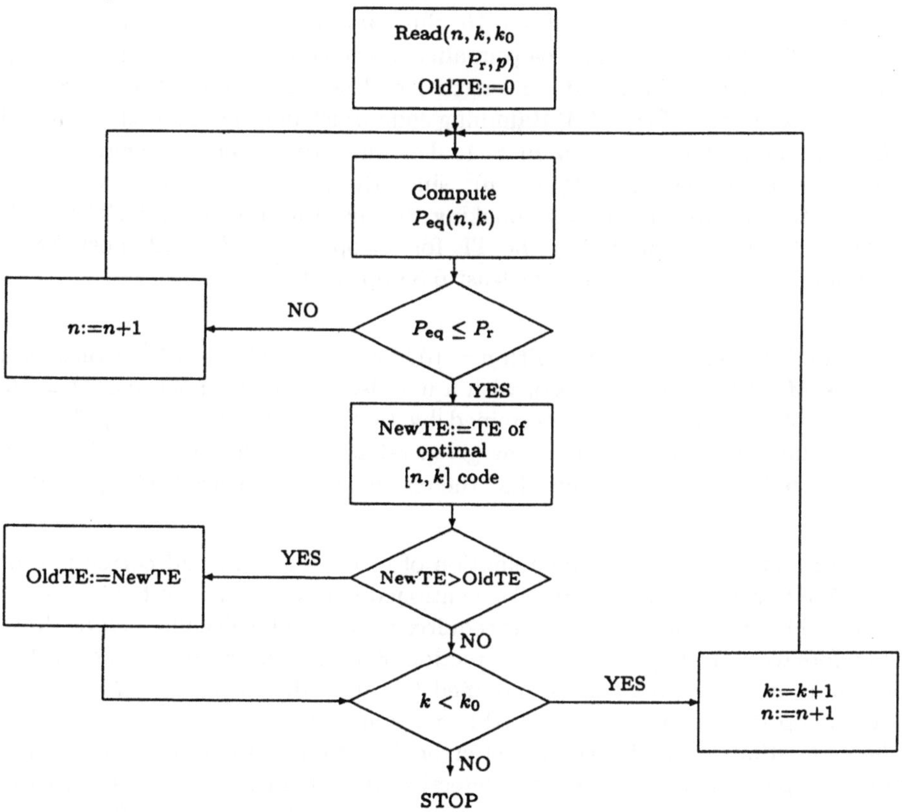

Figure 6.1 Block diagram of the optimization procedure to choose the best code for an ARQ system with the BSC.

It is easy to check that the $[1023, 1013]$ Hamming code satisfies the given reliability and provides TE ≈ 0.98. Therefore the code is at least 0.98-optimal, and there is no need to search further. The code is proper and so $P_{eq} \le 10^{-9}$ for $p \le 10^{-5}$.

Example 2. Consider a moderately noisy BSC with $p = 10^{-3}$. Let us take $P_r = P_{eq} = 10^{-8}$, and use the perfect ARQ protocol with a noiseless return channel. The [63, 51, 5] primitive BCH code gives $P_{eq} < 10^{-8}$ (using the weight distribution of the code) and TE ≈ 0.76. Since $p = 10^{-3} < d/n = 5/63$, we have $P_{eq} < 10^{-8}$ for $p \leq 10^{-3}$. The ϵ-optimality of the code can be estimated using Theoren 6.1. An optimal code of length 7 or less has rate less the $4/7 \approx 0.57$ since this is the rate of the [7, 4] Hamming code which does not have the required reliability, and we can let $f(n,p) = 1$. For an optimal code of length in the interval [8, 15] we get $\eta \leq \frac{11}{15}(1-p)^8$ since the [15, 11] Hamming code does not give the required reliability. Looking through the intervals $[2^m, 2^{m+1} - 1]$ for $m = 4, 5, 6, 7$, we find that the TE for an optimal code is at most 0.887. Therefore the [63, 51, 5] code is at least 0.85-optimal.

Example 3. Let us take a BSC with $p = 10^{-4}$ and $P_r = P_{eq} = 10^{-8}$. Consider a go-back-M ARQ simple protocol with a noiseless return channel and $M = 16$. The TE of this protocol is determined by (5.37) where $\Delta \approx (1-p)^n$. The [63, 57] Hamming code provides the given reliability with TE ≈ 0.821. Using Theorem 6.2.1 we can conclude that the chosen code is at least 0.9-optimal.

Suppose now that we use a combination of error detection and correction for some ARQ protocol on the BSC. As mentioned before we will not try to correct errors of high multiplicity since this requires very complex decoding algorithms. Therefore we restrict ourselves to correction of at most t_0 errors in a block. The procedure in Figure 6.1 can be modified to this end. For each $[n, k]$ code we find the optimal choice of $t \leq t_0$ (this gives an extra loop in the flow chart). The estimation of P_{ue} for combined error detection and correction is slightly more complex than for pure error detection (the main problem in both cases is to determine or estimate the weight distribution of the code). For a rough estimate we can use the bound obtained from (2.12) and Theorem 2.2.11, and for a more precise estimate we use the weight distribution and formula (2.12). An example of a combined correction and detection optimization procedure will be given in the next subsection.

Consider the exceptional case of a pure error detecting code optimization where we can change the channel rate (e.g. time symbol interval) keeping the same source information rate. As a criterion of efficiency we take the coding gain determined by (6.1), where R is changed to the TE η, namely

$$\theta_{(\text{dB})} = 10 \log_{10} \frac{h_{\text{wc}}^2}{h_{\text{c}}^2} \eta. \tag{6.4}$$

Assuming that some modem is used, we have symbol error probability $P(h^2)$ as some function of the signal to noise ratio h^2. Then the value h^2_{wc} in (6.1) can be found from the following equation:

$$P_{\mathrm{r}} = P(h^2_{\mathrm{wc}}). \tag{6.5}$$

We will restrict our consideration to the rather general case of TE represented by (6.2). Then we get

$$\eta = \frac{k}{n} f(n, P(h^2_{\mathrm{c}})). \tag{6.6}$$

The reliability of the information after error detection can for most protocols be determined by (see Chapter 5)

$$P_{\mathrm{r}} = P_{\mathrm{eq}} = \frac{P_{\mathrm{ue}}(C, P(h^2_{\mathrm{c}}))}{k\left(P_{\mathrm{ue}}(C, P(h^2_{\mathrm{c}})) + (1 - P(h^2_{\mathrm{c}}))^n\right)}, \tag{6.7}$$

where C is some $[n,k]$ code.

To calculate the CG for a given reliability P_{r} and given $[n,k]$ code C we should find h^2_{c} satisfying (6.7), put this into (6.6) and then put (6.6) in (6.4). Thus the optimization procedure will be to find the code which provides the highest CG for the given P_{r}. An algorithm for this optimization is similar to the one given in Figure 6.1, but for any parameters (n,k) we should change h^2_{c} to satisfy (6.7) and calculate CG by (6.6) and (6.4).

Example 4. Consider a modem with error probability function

$$P(h^2) = \frac{1}{2} e^{-h^2}, \tag{6.8}$$

this implies a noncoherent optimal receiving of the DPSK binary signals for a channel with additive white gaussian noise. Let the required reliability be $P_{\mathrm{r}} = P_{\mathrm{eq}} = 10^{-7}$, and use the ARQ protocol with perfect selective repeat noiseless return channel. It is easy to check, using (6.4)–(6.8), that the $[63,57]$ Hamming code provides the required reliability and gives CG=2.8 dB. This code may not be optimal, but, on the other hand, we have checked a number of $[63,54]$ codes and the best of these only gave CG=1.8 dB. (It is interesting to note that for pure error correction and unlimited code lengths, one can get CG about 7 dB for $P_{\mathrm{eq}} = 10^{-7}$).

Let us now consider the $q\mathrm{SC}_p$ (without memory). For $n \leq q-1$ there exist Reed-Solomon codes (which are MDS codes and have known weight distributions, see Section 3.10), and we can use these for error detection.

Example 5. Take $q = 16$, $P_r = P_{eq} = 10^{-8}$ and $p = 10^{-2}$. For the perfect selective repeat ARQ protocol with noiseless return channel, we get the $[15, 7; 16]$ RS code as the optimal choice among RS codes, and it provides TE ≈ 0.4.

For real modems q is usually quite small, and at most 8. So the simple choice of Example 4 can not be used in practice. Therefore, consider some modem which provides a $q' = 2^m$-ary symmetric channel without memory. Let $l = ms$ for some integer s, and consider a $q = 2^l$-ary RS code. Its probability of undetected error has the following upper bound which can easily be derived from the known weight distribution of the RS codes:

$$P_{ue} \le (q-1) \sum_{i=d}^{n} \binom{n}{i} \left(\frac{p}{q'-1} (1-p)^{s-1} \right)^i (1-p)^{s(n-i)} \sum_{j=0}^{i-d} (-1)^j \binom{i-1}{j} q^{i-d-j}, \tag{6.9}$$

where p is the symbol error probability provided by the chosen q'-ary modem. Using this bound we can optimize the parameters (n, k, q).

Example 6. Let us have $q = 16$, $q' = 4$, $s = 2$, and the $[15, 8; 16]$ RS code. Using (6.9) we obtain the bounds on $P_{ue}(p)$ shown in Table 6.1 Although the

p	0.001	0.005	0.01	0.05
$P_{ue}(p) \le$	$1.44 \cdot 10^{-23}$	$5.20 \cdot 10^{-18}$	$1.21 \cdot 10^{-15}$	$2.11 \cdot 10^{-10}$
p	0.1	0.2	0.5	0.75
$P_{ue}(p) \le$	$1.96 \cdot 10^{-8}$	$6.69 \cdot 10^{-7}$	$1.63 \cdot 10^{-6}$	$3.73 \cdot 10^{-9}$

Table 6.1 Upper bound on $P_{ue}(p)$ for the $[15, 8; 4]$ RS code with $q = 16$, $q' = 4$, and $s = 2$.

bound (6.9) is not good in the general case, it was strong enough to show that we in the example have the appropriate reliabilities for any channel states. In the particular case when $q' = 2$, the problem is to find the binary weight distribution for a q-ary RS code.

For non-symmetric channels (e.g. the Z-channel) it is not easy to calculate P_{ue} for most codes, and the problem of code optimization is unsolved in general. However, we can use the SATC transform to reduce any non-symmetric binary channel to a BSC and then use the optimization procedure above.

6.3 CODE OPTIMIZATION FOR CHANNELS WITH MEMORY

There are many models for channels with memory (see Chapter 1) and different approaches to estimate $P_{\text{ue}}(C, K)$ for some code C and channel K (see Sections 4.2 and 4.3). Therefore we will describe code optimization for each of these models seperately.

A Composite Channel

In this case the P_{ue} is given by formula (4.16) and if we know the functions $p(\mu)$ and $w(\mu)$ it is no problem to find P_{ue} for any code C with known weight distribution or to estimate it using some upper bound from Chapter 3. However, it is not easy to calculate the integral in (4.16). Furthermore, the function $w(\mu)$ can be known only approximately, and as a rule we need to have a ***robust estimate of reliability.*** Then we can use (5.20) which guarantees the required reliability for any CC with a given average symbol error probability $\bar{p}$ on this channel. Moreover, if $\bar{p}$ is sufficiently small, then $(1-\bar{p})^n \approx 1$ and we do not need to know $\bar{p}$ to give a reliable estimate. The existence of such a robust bound is a very significant advantage of error detection; there is no similar bound for error correcting codes. A very important special case of CC in practice is an ordinary BSC with some given error probability p, but with ***casualty channel states*** when the symbol error probabilities can be close to $\frac{1}{2}$. Information about the the distribution function of these states may be missing and actually it is not so important because the transportations to these states occurs very seldom, and as a rule the channel can be disconnected from the source for some time when the causalty is detected. Then we can use the bound (5.20) letting $\bar{p} = p$ and calculate TE for the ordinary BSC_p. Note that this situation is very typical and can be more useful in practice than a strong BSC. For an $[n, k]$ code C we have

$$P_{\text{wc}}(C, 0, 1/2) \geq P_{\text{ue}}(C, BSC_{1/2}) = \frac{2^k - 1}{2^n} = 2^{-r} - 2^{-n}, \tag{6.10}$$

where $r = n - k$ is the number of check symbols. We have equality in (6.10) if the code C is good or we have used the universal stochastic transform. Assuming this situation we obtain the code optimization procedure shown in Figure 6.2 for $P_{\text{r}} = P_{\text{eq}}$.

The 'good' condition on the code is not necessary if we use a stochastic transform (therefore it is shown in brackets). Otherwise it is necessary to check that

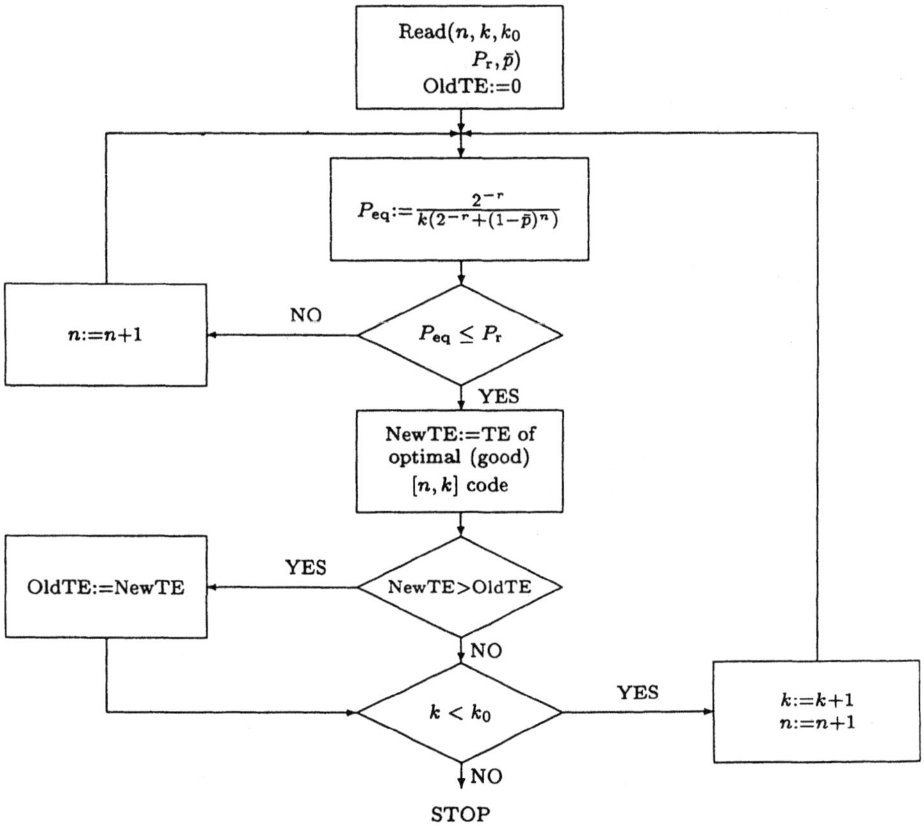

Figure 6.2 Block diagram of the optimization procedure to choose the best robust code for an ARQ system with a CC.

the chosen code is good, using some method described in Chapter 3. Many codes are not good, but experience indicate that the $[n, k, d]$ codes with largest d are more likely to be good, or have the difference between $P_{wc}(C, 0, 1/2)$ and $2^{-r} - 2^{-n}$ small. Note, however, that a large d is no guarantee that the code is good. In fact, examples are known with both d and the ratio $P_{\rm wc}(C)/P_{\rm ue}(C, 1/2)$ arbitrary large, see [99].

Remark. It is interesting that the optimization procedure in Figure 6.2 is simpler than the one in Figure 6.1, because in the last case we do not need to find $P_{\rm ue}$ for different $[n, k]$ codes. This allows us to obtain a fuller list of optimal codes for CC.

The result of some code parameter optimizations for the perfect repeat protocol on the BSC_p with casaulty states, different symbol error probabilities, and different required reliabilities, given as $P_{\rm eq}$, are shown in Table 6.2.

	$P_{\rm r} = 10^{-6}$		$P_{\rm r} = 10^{-8}$		$P_{\rm r} = 10^{-10}$	
$\bar{p}$	(n, k)	η	(n, k)	η	(n, k)	η
0.0001	(340,325)	0.932	(433,415)	0.918	(513,488)	0.904
0.005	(157,144)	0.784	(152,132)	0.746	(178,151)	0.71
0.01	(43,33)	0.412	(61,38)	0.337	(70,41)	0.29
0.05	(31,12)	0.079	(39,13)	0.045	(47,14)	0.027

Table 6.2 The optimal code parameters and the corresponding TE for the perfect selective repeat protocol for different symbol error probabilities $\bar{p}$ and required reliabilities $P_{\rm r}$.

It is not easy to choose a good code for all values of the optimized parameters. Therefore we are forced to apply a stochastic transform to guarantee the required reliability for any casualty state.

Let us now consider a code optimization procedure for an ARQ system when the code is used for combined error detection and correction of errors of multiplicity at most t. We have

$$P_{\rm wc}^{(t)}(C, 0, 1/2) = \max_{0 \le p \le 1/2} P_{\rm ue}^{(t)}(C, BSC_p) \ge \left(2^{-r} - 2^{-n}\right) \sum_{i=0}^{t} \binom{n}{i}, \qquad (6.11)$$

with equallity it the code is t-good. Thus if we choose a t-good code for a CC with casualty channel we get

$$P_{\rm eq} = \frac{\left(2^{-r} - 2^{-n}\right) \sum_{i=0}^{t} \binom{n}{i}}{k \left(\left(2^{-r} - 2^{-n}\right) \sum_{i=0}^{t} \binom{n}{i} + (1 - \bar{p})^n \right)}, \qquad (6.12)$$

where $\bar{p}$ is the average symbol error probability for the given CC.

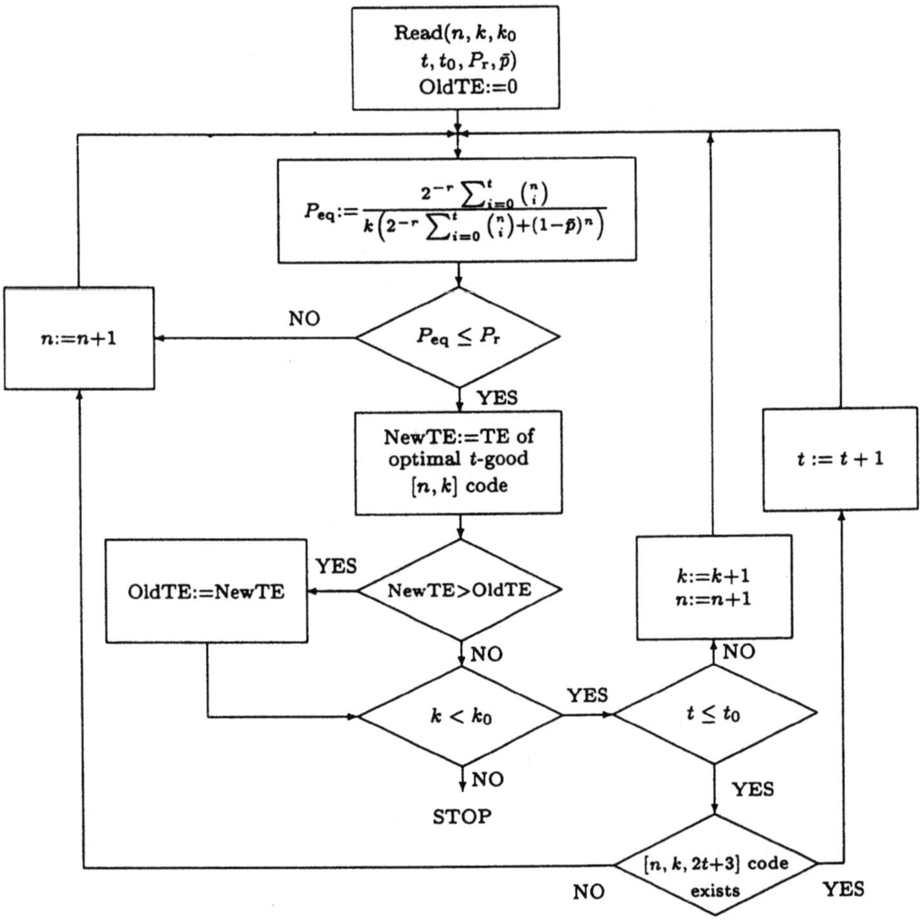

Figure 6.3 Block diagram of the optimization procedure to choose the best robust code for combined error correction and detection for an ARQ system with a CC.

A procedure for code optimization for combined error detection and correction is given in Figure 6.3. We note that it is no point in combining error correction and stochastic transform directly since after SATC and SNTC we obtain a uniform probability distribution of the error patterns and error correction would be useless. However, it has been remarked in 4.2 that we can use two codes, an $[n, k]$ code C for error detection and an $[n', n]$ code C' for error correction; placing the SATC and SMTC between the encoder for C and the encoder for C', and similarly at the reveiving end. An optimization procedure for this construction is given in Figure 6.4. We can see that there is a significant difference between the procedures in Figures 6.3 and 6.4. The procedure in Figure 6.4 is a combination of the procedure in Figure 6.2 and an additional search for an optimal $[n', n]$ code.

The results of optimization of the parameters n, n', k for combined error detection and correction of the perfect repeat protocol on the BSC_p with casulty states, different symbol error probabilities, and different required reliabilities are shown in Table 6.3. Comparing Tables 6.2 and 6.3 we see that combined error detection and correction is superior to error detection only.

	$P_r = 10^{-6}$		$P_r = 10^{-8}$		$P_r = 10^{-10}$	
$\bar{p}$	(n, n', k)	η	(n, n', k)	η	(n, n', k)	η
0.0001	(529,511,499)	0.942	(529,511,493)	0.931	(529,511,486)	0.917
0.001	(383,365,353)	0.873	(389,371,352)	0.856	(416,398,372)	0.840
0.01	(111,97,82)	0.552	(106,92,70)	0.505	(116,102,73)	0.458
0.05	(50,38,19)	0.162	(57,45,19)	0.112	(64,52,19)	0.077

Table 6.3 The optimal code parameters and the corresponding TE for the perfect selective repeat protocol with combined error detection and correction of $t \leq 2$ errors; for different symbol error probabilities $\bar{p}$ and required reliabilities P_r.

Arbitrary Channels With Memory After Stochastic Transforms

Stochastic transforms are used to ensure equality in (6.10) for channels with memory. The optimization procedures shown in Figures 6.2-6.4 for composite channels work also in this case. To calculate TE we use the corresponding expressions for channel models with memory, for example (5.46) for the FPMC and the go-back-M protocol, and (5.46) with $M = 1$ for the perfect simplex selective repeat ARQ protocol, etc.

Channel Models With Memory, but Without Stochastic Transforms

The algorithms in Figures 6.2 and 6.3 can be used in general, provided we can calculate or estimate $P_{ue}(C, K)$ and $P_{ue}^{(t)}(C, K)$ respectively. If it is feasible to

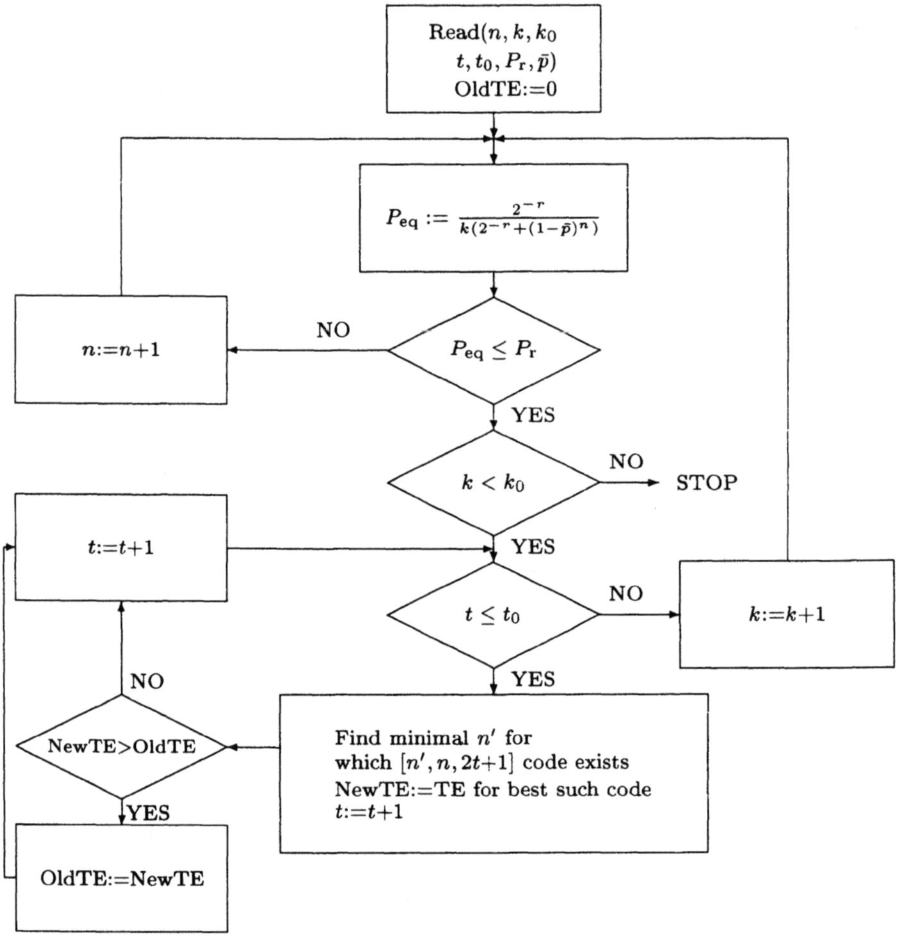

Figure 6.4 Block diagram of the optimization procedure to choose the best robust code for combined error correction and detection for an ARQ system with a CC after stochastic transform.

list the set of code words we can use formulas for the error pattern probabilities to calculate P_{ue} for the given channel model, see (1.14), (1.31), or (1.28), and to choose the best code by exhaustion. In other cases we should use some bounds on P_{ue}, for example (4.12) (or (4.14),(4.15) for combined error detection and correction), where the functions $p(n,i)$ and $P(n,m)$ are taken from the descriptions of the different channel models described in Chapter 1. We obtain then the bounds (4.17), (4.18), (4.19) and an optimization procedure which for given parameters (n,k) choose the $[n,k]$ code, having the lowest value for this bound, in a set of codes with known weight distribution. An exception is the BEC channel model where (4.23) gives an exact expression for $P_{\text{ue}}(C, BEC)$. This relation allows us to carry out a strong optimization procedure (in Figure 6.2) for any function $f(i,n)$ describing the BEC channel and determine a cyclic code with optimized parameters. (Unfortunately, the BEC is an "exotic" model, it does not appear by itself, but only in combination with other models, e.g., the BSC).

For the VPC model with very slowly varying parameters on the block length, we can approximate it by a CC and use the optimization methods described above. Finally, Theorem 4.2.1 shows that for a binary channel with additive noise there exists a code with probability of undeteced error at most 2^{-r}. Therefore we can use the optimization algorithms in Figures 6.3 and 6.4 and for the optimized parameters (n,k) choose a random $[n,k]$ code. We conjecture that such a random code will satisfy (4.3) with significant probability.

Remark. For the LBC-1 model there is the lower bound in Corollary 4.2.4 which says that for any already chosen $[n,k]$ code C there exists a LBC-1 model that gives $P_{\text{ue}}(C, LBC-1) > 2^{-r}$. Also, for a given LBC-1 and given n, k there is no guarantee that there exists an $[n,k]$ code with $P_{\text{ue}}(C, LBC-1) > 2^{-r}$. Therefore, using an $[n,k]$ code chosen at random is bad in the case when some adversary tries to decept us false information, but not in the case when we "play with nature".

6.4 OPTIMIZATION OF ITERATIVE AND CONCATENATED CODES AND PROTOCOLS

For a hybrid ARQ system of type II with a pair of codes (V_1 an $[n,k]$ code and V_2 a $[2k,k]$ code) on BSC with casualty states it is necessary to optimize the parameters (n,k) giving the highest TE by (5.52) and (5.76) for a given reliability determined by equality in (6.10). The procedure in Figure 6.2 works also in this case. A similar approach can be used for hybrid ARQ systems of type II optimizing TE for a given buffer size; in this case we find TE from (5.77).

For the generalized hybrid ARQ scheme, we should optimize the parameters (n, k) of the high rate code and the parameter m of the KM code.

The use of the type II Hybrid-ARQ protocol based on the punctured MDS codes (see Wicker and Bartz [202]) requires optimization of $[n, k]$ MDS code parameters assuming that the code rate of this code is less than $1/2$. Then a procedure similar to the one given in Figure 6.1 is also valid for this case if we change the formula for calculation of the undetected error to the corresponding formula given in [202].

When a type II Hybrid-ARQ protocol is based on the first order Reed-Muller codes (see Wicker and Bartz [203]), the optimization procedure lies in the search of $[n, k]$ Reed-Muller code parameter which give the highest TE, given the required reliability.

An optimization procedure for iterative codes has already been given by Jacobs, see (5.81). The optimization of the recurrent stochastic coding parameters is more complex. The problem is partially solved in [61].

Let us consider in some detail the *optimization problem for ARQ with concatenated codes.* In Chapter 5 we discussed three main ARQ schemes, A, B, and B′. To choose the best A, B, or B′ scheme requires optimization of the inner parameters (n_1, k_1, t) and the outer parameters (n_2, k_2, s_0), for scheme B′ also the parameter α. We will assume that we have a composite channel with casualty states and we try to maximize TE for a given reliability, taking into account the buffer size. If we restrict ourselves to a perfect selective repeat protocol, we can use formula (5.88), where $P_{\text{ue}}(CC)$ is determined by Theorem 4.2.6, to find the reliability for all three strategies, and use formulas (5.89), (5.92) to calculate the TE. The optimization procedure is shown in Figure 6.5.

In the procedure, if $P_{\text{eq}}(CC) > P_{\text{r}}$, then we change the outer parameters; we remark that there always exist some value l which provides a required reliability as can be seen from Theorem (4.2.6) or (4.29). For given k_{10} and k_{20} we can restrict our consideration to lengths n_{10} and n_{20} of the inner and outer codes respectively which have k_{10}/n_{10} and k_{20}/n_{20} less then the TE for some codes already obtained with $n_1 < n_{10}$ and $n_2 < n_{20}$.

In Tables 6.4-6.8 we give the results of some calculations using the procedure for two models of channels and the perfect SR protocol, given a required reliability. For the RFCC it is assumed that all the q-ary symbols of the outer code are transmitted independently.

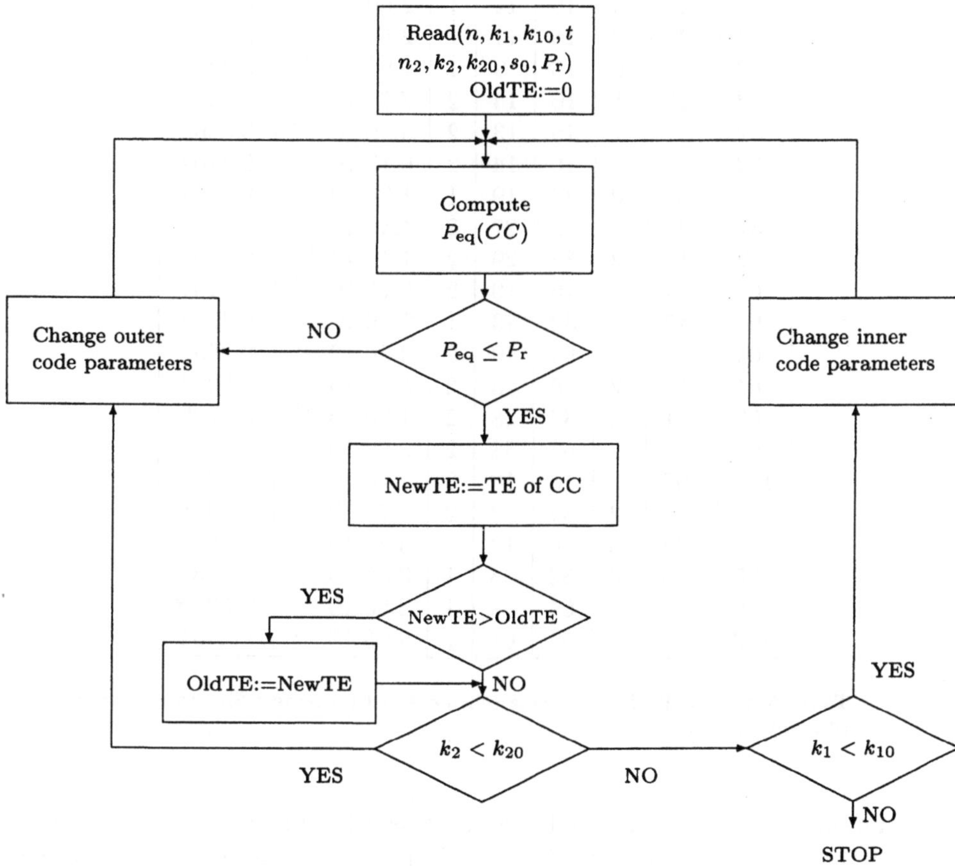

Figure 6.5 Block diagram of the optimization procedure to choose the best CC.

Looking through these tables we can conclude that for BSC with $p = 0.01$, the highest TE is obtained for strategy B for CC with parameters $n_1 = 125$, $k_1 = 96$, $t_1 = 3$, $n_2 = 32$, $k_2 = 31$, $l = 1$. This provides TE ≈ 0.70 and reliability $P_{eq} \approx 7 \cdot 10^{-9}$. For the same channel with $p = 0.05$, strategy B is

n_1	k_1	t_1	n_2	k_2	l	P_{eq}	TE
32	21	0	16	14	1	$2.60{\cdot}10^{-7}$	0.0236
32	21	0	16	11	1	$2.54{\cdot}10^{-8}$	0.2420
32	21	0	16	9	1	$1.90{\cdot}10^{-8}$	0.3219
32	21	1	16	14	2	$7.70{\cdot}10^{-10}$	0.2954
32	21	1	16	13	2	$5.90{\cdot}10^{-10}$	0.3868
32	16	2	16	14	2	$6.17{\cdot}10^{-9}$	0.4104
32	21	0	32	19	1	$1.70{\cdot}10^{-8}$	0.3551
32	21	1	32	27	2	$1.45{\cdot}10^{-9}$	0.3266
32	16	2	32	29	2	$1.50{\cdot}10^{-8}$	0.3465
64	51	1	16	10	2	$6.79{\cdot}10^{-13}$	0.4191
60	47	1	32	23	1	$2.50{\cdot}10^{-12}$	0.4322
64	51	1	64	49	2	$3.30{\cdot}10^{-12}$	0.3557
64	45	2	16	13	2	$5.04{\cdot}10^{-12}$	0.5094
60	41	2	32	28	2	$1.90{\cdot}10^{-11}$	0.5225
64	45	2	64	58	2	$2.20{\cdot}10^{-11}$	0.4866
128	106	2	16	10	2	$7.70{\cdot}10^{-15}$	0.4738
125	103	2	32	23	2	$1.78{\cdot}10^{-14}$	0.5318
128	92	3	16	13	1	$3.40{\cdot}10^{-11}$	0.5695
125	89	3	32	28	1	$3.86{\cdot}10^{-11}$	0.6038
126	90	3	64	57	1	$3.50{\cdot}10^{-11}$	**0.6788**
126	83	4	64	61	1	$9.03{\cdot}10^{-11}$	0.6151

Table 6.4 The TE for CC using scheme A with a perfect SR protocol on a BSC with $p = 0.01$.

again superior to strategy A and it gives TE ≈ 0.39 and reliability $P_{eq} \approx 7{\cdot}10^{-7}$ for CC with parameters $n_1 = 64$, $k_1 = 36$, $t_1 = 4$, $n_2 = 16$, $k_2 = 15$, $l = 1$. The optimal CC requires correction of 3 or 4 errors by the inner code. If we restrict our consideration to CC which correct at most two errors we get TE ≈ 0.6 and $P_{eq} \approx 1.2 \cdot 10^{-7}$ for $p = 0.01$ and CC with parameters $n_1 = 64$, $k_1 = 45$, $t_1 = 2$, $n_2 = 16$, $k_2 = 15$; and TE ≈ 0.3 and $P_{eq} \approx 6.8 \cdot 10^{-9}$ for $p = 0.05$ and $n_1 = 32$, $k_1 = 10$, $t_1 = 2$, $n_2 = 16$, $k_2 = 14$, $l = 2$; in both cases for strategy B.

Strategy B is also superior for RCF, for $h^2 = 98$ ($p = 0.01$) it gives TE ≈ 0.78 and reliability $P_{eq} \approx 3 \cdot 10^{-13}$ for CC with parameters $n_1 = 128$, $k_1 = 111$, $t_1 = 0$, $n_2 = 64$, $k_2 = 63$; for $h^2 = 18$ ($p = 0.05$) it gives TE ≈ 0.50, $P_{eq} \approx 3.3 \cdot 10^{-13}$ for CC with parameters $n_1 = 125$, $k_1 = 110$, $t_1 = 0$, $n_2 = 32$, $k_2 = 31$, $l = 1$.

As mentioned in Chapter 5, strategy B requires much more buffer space than stategy A. This is particularly significant for channels with large round-trip-

n_1	k_1	t_1	n_2	k_2	l	P_{eq}	TE
32	16	2	16	12	2	$2.37 \cdot 10^{-8}$	0.1070
32	16	2	16	9	2	$1.13 \cdot 10^{-8}$	0.2237
32	16	2	32	24	2	$3.60 \cdot 10^{-8}$	0.1427
32	16	2	32	20	2	$2.29 \cdot 10^{-8}$	0.2286
64	36	3	16	10	1	$3.02 \cdot 10^{-8}$	0.1169
64	36	3	16	7	1	$1.67 \cdot 10^{-8}$	0.2112
64	30	4	16	12	1	$5.07 \cdot 10^{-8}$	0.1883
64	30	4	16	10	1	$3.60 \cdot 10^{-8}$	0.2606
60	26	4	32	28	1	$2.86 \cdot 10^{-7}$	0.0554
60	26	4	32	22	1	$5.61 \cdot 10^{-8}$	0.2828
60	26	4	64	50	1	$5.49 \cdot 10^{-8}$	0.1736
60	26	4	64	49	1	$4.48 \cdot 10^{-8}$	0.2130
60	26	4	64	48	1	$3.86 \cdot 10^{-8}$	0.2471
60	26	4	64	47	1	$3.49 \cdot 10^{-8}$	0.2736
60	26	4	64	46	1	$3.27 \cdot 10^{-8}$	0.2914
60	26	4	64	45	1	$3.15 \cdot 10^{-8}$	0.2952
60	26	4	64	44	1	$3.11 \cdot 10^{-8}$	0.3024
60	26	4	64	43	1	$3.11 \cdot 10^{-8}$	**0.3064**

Table 6.5 The TE for CC using scheme A with a perfect SR protocol on a BSC with $p = 0.05$.

delays. Strategy B′ allows for a tradeoff between TE and buffer size, and occupies an intermediate position between A and B. The optimization procedure can be modified to take into account the buffer size.

6.5 SOME CONSIDERATIONS ABOUT OPTIMIZATION OF PROTOCOLS

It is difficult to make a theory of protocol optimization, because each protocol is appropriate for some channel and its implementation depends on restrictions on the complexity of hardware or software. For a channel with half duplex regime it is natural to implement the stop-and-wait protocol. If there are very hard restrictions on the complexity of hardware or software and simultaneously the forward and return channels are moderately noisy and the round-trip-delay is very small, we could prefer the go-back-M protocol for a duplex system. The trivially optimal protocol is the perfect selective repeat protocol since this has

a) $p = 0.01$

n_1	k_1	t_1	n_2	k_2	l	P_{eq}	TE
32	21	0	16	15	1	$1.01 \cdot 10^{-8}$	0.4416
32	16	1	16	15	1	$1.17 \cdot 10^{-7}$	0.4462
64	51	0	16	15	1	$4.72 \cdot 10^{-11}$	0.3913
64	45	2	16	15	1	$1.18 \cdot 10^{-7}$	0.6037
60	47	1	32	30	2	$1.93 \cdot 10^{-8}$	0.4968
60	36	2	32	31	1	$3.09 \cdot 10^{-9}$	0.5663
128	106	1	16	15	1	$2.11 \cdot 10^{-12}$	0.4914
128	99	3	16	15	1	$6.17 \cdot 10^{-9}$	0.6888
125	103	1	32	31	1	$2.43 \cdot 10^{-12}$	0.5135
125	96	3	32	31	1	$6.87 \cdot 10^{-9}$	**0.7032**
126	97	3	64	63	1	$6.61 \cdot 10^{-9}$	0.6997

b) $p = 0.05$

n_1	k_1	t_1	n_2	k_2	l	P_{eq}	TE
32	21	0	16	15	1	$1.04 \cdot 10^{-8}$	0.1148
32	21	1	16	14	2	$6.51 \cdot 10^{-10}$	0.1889
32	10	2	16	14	2	$6.82 \cdot 10^{-9}$	0.2951
64	51	0	16	15	1	$4.96 \cdot 10^{-11}$	0.0267
64	51	1	16	14	2	$8.69 \cdot 10^{-13}$	0.0562
64	45	2	16	15	1	$1.31 \cdot 10^{-7}$	0.2101
64	39	3	16	15	1	$3.15 \cdot 10^{-7}$	0.3326
64	36	4	16	15	1	$7.14 \cdot 10^{-7}$	**0.3941**
64	36	4	16	14	2	$1.31 \cdot 10^{-10}$	0.3678

Table 6.6 The TE for CC using scheme B with a perfect SR protocol on a BSC.

the highest TE for a given reliability; however it only works when the return channel is noiseless. If we have a duplex channel with large round-trip-delay and high noise (for example a satellite channel), the protocol with address confirmation or the SR-protocol with finite buffer size are the best. However, they require a rather complex control algorithm and buffers. If we want as high TE as possible, it is reasonable to use different modifications of hybrid protocols of type II. But in some cases "the game will not be worth the candle" unless the channel is a mixture of BSC and bursts of errors and the code parameters are matched to the burst lengths. The main problem will be to find an error correcting algorithm with a moderate complexity. Therefore KM codes are very usefull in this case.

a) $h^2 = 98$ ($\bar{p} = 0.01$)

n_1	k_1	t_1	n_2	k_2	l	P_{eq}	TE
32	26	0	16	13	1	$3.70 \cdot 10^{-7}$	0.4897
64	57	0	16	12	1	$8.63 \cdot 10^{-9}$	0.5750
60	53	0	32	26	1	$1.27 \cdot 10^{-8}$	0.5457
64	57	0	64	54	1	$1.11 \cdot 10^{-8}$	0.4479
128	120	0	16	12	1	$2.31 \cdot 10^{-10}$	0.6296
126	116	0	64	53	1	$2.46 \cdot 10^{-10}$	0.5200
128	120	0	128	109	1	$3.06 \cdot 10^{-10}$	0.4753
128	113	0	16	12	1	$3.78 \cdot 10^{-13}$	0.6248
125	110	0	32	25	1	$3.52 \cdot 10^{-13}$	0.6726
126	111	0	64	53	1	$3.35 \cdot 10^{-13}$	0.7074
128	113	0	128	108	1	$3.24 \cdot 10^{-13}$	**0.7322**

b) $h^2 = 18$ ($\bar{p} = 0.05$)

n_1	k_1	t_1	n_2	k_2	l	P_{eq}	TE
32	21	0	16	14	1	$4.54 \cdot 10^{-7}$	0.0135
32	21	0	16	8	1	$2.05 \cdot 10^{-8}$	0.2991
30	19	0	32	18	1	$1.83 \cdot 10^{-8}$	0.3349
64	51	0	16	8	1	$1.07 \cdot 10^{-7}$	0.3289
60	47	0	32	26	1	$4.85 \cdot 10^{-9}$	0.0072
60	47	0	32	16	1	$9.50 \cdot 10^{-11}$	0.3702
60	47	0	64	34	1	$8.74 \cdot 10^{-11}$	0.4028
125	110	0	32	18	1	$8.14 \cdot 10^{-13}$	0.2912
125	110	0	32	15	1	$6.37 \cdot 10^{-13}$	0.3723
125	111	0	64	32	1	$6.06 \cdot 10^{-13}$	**0.4075**

Table 6.7 The TE for CC using scheme A with a perfect SR protocol on a RFCC.

Recurrent stochastic coding used together with a SR finite buffer size is irreplaceable for channels with memory when we need both guaranteed reliability and maximum TE.

Concatenated codes used with strategies A, B, or B′ together with a SR-finite size buffer protocol relieves us the task of finding a good code. They are irreplaceable in cases when we use multitone modems which bring us to the local binomial channel model. For all these protocols we can optimize the code parameters using the procedures described above.

a) $h^2 = 98$ ($\bar{p} = 0.01$)

n_1	k_1	t_1	n_2	k_2	l	P_{eq}	TE
32	26	0	16	15	1	$3.11{\cdot}10^{-7}$	0.5547
32	26	0	16	14	2	$1.45{\cdot}10^{-11}$	0.5177
32	21	0	16	15	1	$1.08{\cdot}10^{-8}$	0.5701
32	21	0	16	14	2	$1.59{\cdot}10^{-14}$	0.5321
30	24	0	32	31	1	$5.59{\cdot}10^{-7}$	0.4336
30	19	0	32	31	1	$5.60{\cdot}10^{-7}$	0.5645
64	51	0	16	15	1	$4.72{\cdot}10^{-11}$	0.6867
60	47	0	64	63	1	$7.37{\cdot}10^{-11}$	0.7052
128	111	0	64	63	1	$3.06{\cdot}10^{-13}$	**0.7861**

b) $h^2 = 18$ ($\bar{p} = 0.05$)

n_1	k_1	t_1	n_2	k_2	l	P_{eq}	TE
32	21	0	16	15	1	$1.01{\cdot}10^{-8}$	0.4228
30	19	0	32	31	1	$1.61{\cdot}10^{-8}$	0.4200
64	51	0	16	15	1	$4.72{\cdot}10^{-11}$	0.4795
60	47	0	32	31	1	$7.47{\cdot}10^{-11}$	0.4891
60	47	0	64	63	1	$7.40{\cdot}10^{-11}$	0.4940
128	113	0	16	15	1	$2.87{\cdot}10^{-13}$	0.4931
125	110	0	32	31	1	$3.30{\cdot}10^{-13}$	**0.5088**
126	111	0	64	63	1	$2.78{\cdot}10^{-13}$	0.5086

Table 6.8 The TE for CC using scheme B with a perfect SR protocol on a RFCC.

6.6 COMMENTS AND REFERENCES

This chapter is more practical then the previous ones with emphesis on practical procedures. It is possible that further advances in the theory of error detecting codes can bring us further practical tools, but we feel that the present methods can provide good solutions in most cases.

The best understood error detecting code optimization problem is the case of composite channels (in particular BSC with casualty), because the problem can be split into two subproblems: optimization of the parameters (n, k, t_0) and search for *good* or *t-good* codes (it we use stochastic transforms, only the first subproblem has to be solved). The first subproblem can be solved by the exhaustive search procedures described in Section 6.3. (It would be possible to obtain an explicit solution for this problem for a given protocol, but the procedures will give a solution in a few second on a moderate computer, even for block lengths up to tens of thousands).

The solution of the second subproblem is not trivial even though we have criteria to check if some given code is good or t-good. In practice it is feasable only if the dimension or redundancy of the code is less than 30-40, if the weight distribution has to be determined by a complete listing of all code words. (However, this is sufficent if the required reliability is of the order 10^{-9}).

As shown in Chapter 3, the ratio $P_{\text{wc}}(C, 0, \frac{1}{2})/P_{\text{ue}}(C, \frac{1}{2})$ is small even for most codes which are not good, and the code may therefore be good enough for many practical application.

A second method to remove the second subproblem is to use concatenated codes. This approach also results in an increase in the TE for a given reliability. The drawback is that they require large buffers, and so some effective selective repeat or address confirmation protocols are needed.

We have not given error detecting code optimization procedures for hybrid protocol of type II. But it is doubtful that in this case one can obtain significant advances over the concatenated protocols for the same complexity of hardware or software.

The results given in this chapter are mostly new. The algorithm for concatenated code optimization is from Korzhik et.al. [119]. The example of optimization for concatenated code parameters for schemes A, B, and B′ is due to Malev [143]. He also investigated so-called adaptive ARQ systems, where combined error detection and correction can be choosen depending on the state of the channel. An adaptive hybrid ARQ scheme for slowly time-varying channels has been considered by Memişoğlu and Bilgen [147].

Another approach to an adaptive error control system using hybrid ARQ schemes was given by Kousa and Rahman [122].

The optimization of Hybrid-ARQ protocol based on the punctured MDS codes or first-order Reed-Muller codes can be done using the results of Wicker and Bartz [202], [203].

7

CONCLUDING REMARKS AND SOME OPEN ENDS

In this book we have given a comprehensive treatment of the error detecting properties of block codes, in particular binary linear codes. Their main application is in communication systems with feedback, and a large part of the book is devoted to ARQ protocols and optimization of throughput efficiency.

An area which is outside the scope of this book is the use of convolutional codes. For pure error detection they offer no advantage over block codes, but for combined error detection and correction using soft decision they may be a good alternative.

There are a large number of open problems connected with the area of error detection covered in the book. Below we list some (research) problems and problem areas where more study is needed or desired. Some of them may be easy, other very hard, some are general, other more specific.

1. Determine the weight distribution for further classes of codes. Determine the generalized weight distribution defined on page 30 for some central classes of codes.

2. For all binary codes we have considered, the function $P_{\text{ue}}(C,p)$ has at most one maximum on the open interval $(0,1/2)$. Decide if this is true for any codes (or for any linear codes). Is it always true that $P_{\text{ue}}(C,p)$ has at most two maxima on the closed interval $[0,1]$?

3. Find better algorithms to decide if a linear code is good or t-good for some given t. Determine the complexity of this problem in general. Will knowledge of the minimum distance of C_G change the complexity?

4. Determine optimal codes for further sets of parameters.
5. Improve the lower and upper bounds on P_{ue}.
6. Find improved upper and lower bounds on the worst-case error probability of a linear code on the BSC.
7. Characterize the class of $[n, k; q]$ codes such that $P_{\text{wc}}(C, 0, 1) = (q^k - 1)/q^n$ (for the qSC), or find classes of such codes.
8. Improve the bound in Theorem 3.8.3.
9. Find a closed formula for $P_{\text{ue}}(C)$ defined by (4.21) for some classes of codes.
10. Derive an expression for the TE of the selective-repeat + go-back-M protocol on the FPMS channel model.
11. Generalize to $M > 2$ the generating function $f_{IO}(x)$ given by (5.66) for $M = 2$.
12. Construct an iterative protocol, with the property $P_{\text{fd}} \to 0$ for $n \to \infty$, $R \geq R_0 > 0$, which is proper for a noisy return channel.
13. Construct protocols with information feedback which require finite buffer and are valid on a noisy return channel.
14. Develop an optimization algorithm for the hybrid ARQ protocols of type II similar to those given in Figures 6.1-6.4.
15. Find codes and protocols which give significantly better results than those given by the concatenated codes and protocols referred in Tables 6.4-6.8.
16. Develop an algorithm for combined protocol and code optimization given a channel, a required reliability, and restrictions on buffer size and complexity.

REFERENCES

[1] Lars Alfors, *Complex Analysis*, McGraw-Hill Book Company, New York, 1953.

[2] D. Bertsekas and R. Gallager, *Data networks*, Prentice-Hall, Inc., Englewood Cliffs, New Jersey, 1986.

[3] Birkhoff and G. MacLane, *A Survey of Modern Algebra*, The MacMillan Company, New York, 1953.

[4] M. Blaum, *Codes for detecting and correcting unidirectional errors*, IEEE Comp. Soc. Press, Los Alamitos, Cal. 1993.

[5] G. C. Clark and J. B. Cain, *Error Correcting Codes for Digital Communication*, Plenum Press, New York 1981.

[6] W. Feller, *An introduction to probability theory and its applications*, 2nd ed., vol.1, John Wiley & Sons, Inc., London 1957.

[7] L. M. Fink, *Theory of Digital Communication* (in Russian), Sov. Radio, Moscow 1965.

[8] E. Isaacson and H. B. Keller, *Analysis of Numerical Methods*, John Wiley & Sons, Inc., New York, 1966.

[9] Z. Kanevsky, *Transmission of Messages by Information Feedback Systems* (in Russian), Svyaz, Moscow 1969.

[10] V. I. Korzhik and L. M. Fink, *Noise-Stable Coding of Discrete Messages in Channels with a Random Structure*, (in Russian), Svyaz, Moscow 1975.

[11] S. Lin and D. J. Costello, *Error Control Coding, Fundamentals and Applications*, Prentice-Hall Publ, Englewood Cliffs, New Jersey, 1983.

[12] J. H. van Lint, *Introduction to Coding Theory*, Springer-Verlag, New York 1982.

[13] J. MacWilliams and N. J. A. Sloane, *The theory of Error-Correcting Codes*, North-Holland Publishing Company, Amsterdam 1977.

[14] Ø. Ore, *Theory of Graphs*, Amer. Math. Soc., Providence 1962

[15] W. W. Peterson, *Error-Correcting Codes*, MIT Press, Cambridge, Mass. 1961.

[16] W. W. Peterson and E. J. Weldon, *Error-Correcting Codes*, MIT Press, Cambridge, Mass. 1972.

[17] A. Poli and Ll. Huguet, *Error Correcting Codes, Theory and Applications*, Prentice Hall UK, Hemel Hempstead 1992.

[18] J. Proakis, *Digital Communication*, McGraw-Hill Book Company, New York.

[19] L. Purtov (ed.), *Elements of Digital Information Transmission Theory* (in Russian), Svyaz, Moscow 1972.

[20] T.R.N. Rao and E. Fujiwara, *Error-control coding for computer systems*, Prentice Hall, Inc., Englewood Cliffs, New Jersey, 1989.

[21] W. Rudin, *Real and Complex Analysis*, McGraw-Hill Book Company, New York, 1970.

[22] V. I. Turin, *Information Transmission in Channels With Memory* (in Russian), Svyaz, Moscow 1977.

[23] H. Zimmermann and S. Mason, *Electronic Circuits, Signals and Systems*, John Wiley & Sons, Inc., New York, 1960.

PAPERS.

[24] V. K. Agarwal and A. Ivanov, "Computing the probability of undetected error for shortened cyclic codes", *IEEE Trans. Commun.* 40 (1992) 494-499.

[25] K.A.S. Abdel-Ghaffar, "A lower bound on the undetected probability of block codes", to be presented at the 1995 IEEE Int. Symp. Inform. Theory.

[26] A. Amosov and N. Aleksandrov, "Data transmission with address confirmation" (in Russian), *Electrosvyaz* No. 4 (1974).

[27] T. C. Ancheta, "An upper bound on the ratio of the probabilities of subgroups and cosets", *IEEE Trans. Inform. Theory* 27 (1981) 646-647.

[28] E. F. Assmus and H. F. Mattson, "The weight distribution of a coset of a linear code", *IEEE Trans. Inform. Theory* 24 (1978) 497.

[29] A. M. Barg and I. I. Dumer, "On computing the weight spectrum of cyclic codes", *IEEE Trans. Inform. Theory* 38 (1992) 1382-1386.

[30] G. Benelli, "A new method for integration of modulation and channel coding in an ARQ protocol", *IEEE Trans. Commun.*, 40 (1992) 1594-1606.

[31] G. Benelli, "Some ARQ protocols with finite receiver buffer", *IEEE Trans. Commun.*, 41 (1993) 513-523.

[32] G. Benelli, "New ARQ protocols using concatenated codes", *IEEE Trans. Commun.*, 41 (1993) 1013-1019.

[33] G. Benelli, "A new selective ARQ protocol with finite length buffer", *IEEE Trans. Commun.*, 41 (1993) 1102-1111.

[34] G. Benelli and A. Garzelli, "Type-II hybrid ARQ protocol using concatenated codes", *IEE Proc.-I Commun., Speech and Vision*, 140 (1993) 346-350.

[35] R. J. Benice and A. H. Frey Jr., "An analysis of retransmission systems", *IEEE Trans. Commun.*, 12 (1964) 135-145.

[36] J. M. Berger, "A note on error detecting codes for asymmetric channels", *Inform. Control* 4 (1961) 68-73. Reprinted in [4].

[37] E. R. Berlekamp, R. J. McEliece, H. C. A. van Tilborg, "On the inherent intractability of certain coding problems", *IEEE Trans. Inform. Theory* 24 (1978) 384-386.

[38] V. Blinovsky, "New estimation of the probability of undetected error", to be presented at the 1995 IEEE Int. Symp. Inform. Theory.

[39] I. Blake and K. Kith, "On the complete weight enumerator of Reed-Solomon codes" *SIAM Journal on Discrete Math.* 4 (1991) 164-171.

[40] M. Blaum, J. Bruck, and L. Tolhuizen, "A note on "A systematic (12,8) code for correcting single errors and detecting adjacent errors"", *IEEE Trans. Comput.* 43 (1994) 125.

[41] I. E. Bocharova and B. D. Kudryashov, "Development of a discrete channel model for fading channels" *Prob. Peredachi Inform.* 29 no.1 (1993) 58-67 (in Russian) [English translation: *Prob. Inform. Transmission* 29 (1993) 50-57].

[42] J. M. Borden, "Optimal asymmetric error detecting codes", *Inform. Control* 53 (1982) 66-73. Reprinted in [4].

[43] P. Camion, B. Courteau, and A. Montpetit, "Weight distribution of cosets of 2-error-correcting binary BCH codes of length 15, 63, and 255, *IEEE Trans. Inform. Theory* 38 (1992) 1353-1357.

[44] J. L. Carter and M. N. Wegman, "Universal classes of hash functions", *J. Comp. System Sci.* 18 (1979) 143-154.

[45] G. Castagnoli, J. Ganz, and P. Graber, "Optimal cyclic redundancy-check codes with 16 bit redundancy", *IEEE Trans. Commun.* 38 (1990) 111-114.

[46] G. Castagnoli, S. Bräuer, and M. Herrmann, "Optimization of cyclic redundancy-check codes with 24 and 32 parity bits", *IEEE Trans. Commun.* 41 (1993) 883-892.

[47] E. H. Chang, "Theory of information feedback systems", *IRE Trans. Inform. Theory* 2, no. 3 (1956) 29-42.

[48] E. H. Chang, "Improvement of two-way communication by means of feedback", *IRE Nat. Conv. Rec* pt 4 (1966) 88-104.

[49] P. Charpin, "Weight distribution of the cosets of the binary 2-error-correcting BCH codes"' *C. R. Acad. Sci. I, Math.* 317, no. 10 (1994) 975-980.

[50] D. Chun and J. K. Wolf, "Special hardware for computing the probability of undetected error for certain binary CRC codes and test results", *IEEE Trans. Commun.* 42 (1994) 2769-2772.

[51] K. M. Cheung, "Identities and approximations for the weight distribution of *Q*-aey codes", *IEEE Trans. Inform. Theory* 36 (1990) 1149-1153.

[52] M. A. de Boer, "Almost MDS codes", preprint (1994).

[53] P. Delsarte, "Bounds for unrestricted codes, by linear programming", *Philips Res. Reports* 27 (1972) 272-289.

[54] R. H. Deng, "Hybrid ARQ schemes for point-to-multipoint communication over nonstationary broadcast channels", *IEEE Trans. Commun.* 41 (1993) 1379-1387.

[55] R. L. Dobrushin, "The mathematical problems of Shannon's theory of optimal information coding" (in Russian), *Prob. Peredachi Inform.* 10, 1961, 63-107.

[56] S. M. Dodunekov and I. N. Landgev, "On near-MDS codes", Report LiTH-ISY-R-1563, Dept. of Elec. Eng., Linköping Univ. (1994).

[57] T. Etzion, "Optimal codes for correcting single errors and detecting adjacent errors", *IEEE Trans. Inform. Theory* 38 (1992) 1357-1360.

[58] A. Faldum and W. Willems, "Codes of small defect", manuscript 1995.

[59] R. Fantacci, "Performance evaluation of some efficient stop-and-wait techniques", *IEEE Trans. Commmun.* 40 (1992) 1665-1669.

[60] L. M. Fink and V. I. Korzhik, "Selection of codes for error detection in decision-feedback systems" (in Russian), in *Proc. Third Int. Symp. Inform. Theory* no. 1 (1973) 138-142.

[61] L. M. Fink and S. A. Mukhametshina, "Pseudostochastic coding for error detection", *Prob. Peredachi Inform.* 15, no. 2 (1979) 36-39 (in Russian) [English translation: *Prob. Inform. Transmission* 15 (1979) 108-110].

[62] S. Fratini, "Error detection in a class of decimal codes", *IEEE Trans. Inform. Theory* 35 (1989) 1095-1098.

[63] C. V. Freiman, "Optimal error detection codes for completely asymmetric channels", *Inform. Control* 5 (1962) 64-71. Reprinted in [4].

[64] E. Fujiwara and M. Sakura, "Nonsystematic d-unidirectional error detecting codes", *IEEE Int. Symp. Inform. Theory* (1990) 174.

[65] T. Fujiwara and T. Kasami, "Probability of undetected error after decoding for a concatenated coding scheme", *IEEE Int. Symp. Inform. Theory* (1985) 110.

[66] T. Fujiwara and T. Kasami, "Error detecting capabilities of the shortened Hamming codes adopted for error detection in IEEE standard 802.3", *IEEE Int. Symp. Inform. Theory* (1986) 65.

[67] T. Fujiwara, T. Kasami, and S. Feng, "On the monotonic property of the probability of undetected error for a shortened code", *IEEE Trans. Inform. Theory* 37 (1991) 1409-1411.

[68] T. Fujiwara, T. Kasami, A. Kitai, and S. Lin, "On the undetected error probability for shortened Hamming codes", *IEEE Trans. Commun.* 33 (1985) 570-574.

[69] T. Fujiwara, T. Kasami, and S. Lin, "Error detecting capabilities of the shortened Hamming codes adopted for error detection in IEEE standard 802.3", *IEEE Trans. Commun.* 37 (1989) 986-989.

[70] T. Fujiwara, A. Kitai, S. Yamamura, T. Kasami, and S. Lin, "On the undetected error probability for shortened cyclic Hamming codes", in *Proc. 5th Conf. Inform. Theory and Its Appl.*, Hachimantai, Japan, Oct. 1982.

[71] G. Funk, "Message error detection properties of HDLC protocols", *IEEE Trans. Commun.* 30 (1982) 252-257.

[72] G. Funk, "Determination of best shortened linear codes", manuscript.

[73] S. W. Golomb, "A probability calculation and estimate for Reed-Solomon codes", unpublished.

[74] T. A. Gulliver and V. K. Bhargava, "A systematic (16,8) code for correcting double errors and detecting triple-adjacent errors", *IEEE Trans. on Computers* 42 (1993) 109-112.

[75] Y. Hayashida, "Throughput analysis of tandem-type go-back-N ARQ scheme for satellite communication", *IEEE Trans. Commun.* 41 (1993) 1517-1524.

[76] A. R. Hammons, P. V. Kumar, A. R. Calderbank, N. J. A. Sloane, and P. Solé, "The Z_4-linearity of Kerdock, Preparata, Goethals and related codes", *IEEE Trans. Inform. Theory* 40 (1994) 301-319.

[77] B. C. Harvey and S. B. Wicker, "Packet combining systems based on the Viterbi decoder", *IEEE Trans. Commmun.* 42 (1994) 1544-1557.

[78] T. Hashimoto, "Good error detection codes satisfy the expurgated bound", to be presented at the 1995 IEEE Int. Symp. Inform. Theory.

[79] T. Helleseth, T. Kløve, V. Levenshtein, Ø. Ytrehus, "Bounds on the minimum support weights", *IEEE Trans. Inform. Theory* 41 (1995) 432-440.

[80] T. Helleseth, T. Kløve, J. Mykkeltveit, "The weight distribution of irreducible cyclic codes", *Discrete Math.* 18 (1977) 179-211.

[81] T. Helleseth, T. Kløve, Ø. Ytrehus, "Generalized Hamming weights of linear codes", *IEEE Trans. Inform. Theory*, vol. 38, 1133-1140, 1992.

[82] T. Helleseth, T. Kløve, Ø. Ytrehus, "On generalizations of the Griesmer bound", *Reports in informatics*, no. 87, Department of Informatics, University of Bergen, September 1993.

[83] M. E. Hellman, "Error detection made simple", in *IEEE Int. Conf. on Commun.*, Minneapolis, June 17-19 1974, 9A1-9A4.

[84] W. H. Huggins, "Signal-flow graphs and random signals", *Proc. IRE* January (1957) 74-86.

[85] Z. Huntoon and A. M. Michelson, "On the computation of the probability of post-decoding error events for block codes", *IEEE Trans. Inform. Theory* 23 (1977) 399-403.

[86] K. Imamura, K. Tokiwa, M. Kasahara, "On computation of the binary weight distribution of some Reed-Solomon codes and their extended codes", *Proc. Int. Colloq. Coding Theory*, Osaka, Japan (1988) 195-204.

[87] I. Jacobs, "Optimum error detections codes for noiseless decision feedback", *IRE Trans. Inform. Theory* 8 (1962) 369-371.

[88] R. Johansson, "A class of (12,8) codes for correcting single errors and detecting double errors within a nibble", *IEEE Trans. Comput.* 42 (1993) 1504-1506.

[89] M. G. Karpovsky and P. Nagvajara, "Asymptotically bent functions and optimal quadratic codes for minimax criterion on error detection", *IEEE Trans. Inform. Theory* 35 (1989) 1299-1305.

[90] T. Kasami, "The weight enumerators for several classes of subcodes of the 2nd order binary Reed-Muller codes", *Inform. Control* 18 (1971) 369-394.

[91] T. Kasami, T. Fujiwara, S. Lin, "Generalized Sidel'nikov's bound and its applications", *IEEE Int. Symp. Inform. Theory* (1983) 120.

[92] T. Kasami, T. Fujiwara, S. Lin, "An approximation to the weight distribution of binary linear codes", *IEEE Trans. Inform. Theory* 31 (1985).

[93] T. Kasami, T. Fujiwara, S. Lin, "A concatenated coding scheme for error control", *IEEE Trans. Commun.* 34 (1986) 481-488.

[94] T. Kasami, T. Kløve, S. Lin, "Linear block codes for error detection", *IEEE Trans. Inform. Theory* 29 (1983) 131-136.

[95] T. Kasami and S. Lin, "On the probability of undetected error for the maximum distance separable codes", *IEEE Trans. Commun.* 32 (1984) 998-1006.

[96] T. Kasami and S. Lin, "The binary weight distribution of the extended $(2^m, 2^m - 4)$ code of the Reed-Solomon codes over $GF(2^m)$ with generator polynomial $(x - \alpha)(x - \alpha^2)(x - \alpha^3)$", *Linear Algebra and Its Applications* 98 (1988) 291-307.

[97] G. L. Katsman, "Upper bounds on the probability of undetected error", *Proc. Fourth Intern. Workshop on Algebraic and Comb. Coding Th.*, Novgorod, Russia, Sept. 11-17 (1994).

[98] S. R. Kim and C. K. Un, "Throughput analysis for two ARQ schemes using combined transition matrix", *IEEE Trans. Commmun.* 40 (1992) 1679-1683.

[99] T. Kløve, "The probability of undetected error when a code is used for error correction and detection", *IEEE Trans. Inform. Theory* 30 (1984) 388-392.

[100] T. Kløve, "Generalizations of the Korzhik bound", *IEEE Trans. Inform. Theory* 30 (1984) 771-773.

[101] T. Kløve, "Using codes for error correction and detection", *IEEE Trans. Inform. Theory* 30 (1984) 868-871.

[102] T. Kløve, "Codes for error detection", unpublished lecture notes, presented in the Nordic Summer School on Coding Theory, Bergen 1987.

[103] T. Kløve, "Optimal codes for error detection", *IEEE Trans. Inform. Theory* 38 (1992) 479-489.

[104] T. Kløve, "The weight distribution of cosets", *IEEE Trans. Inform. Theory, IEEE Trans. Inform. Theory* 40 (1994) 911-913.

[105] T. Kløve, "A lower bound on the probability of undetected error", *Proc. Sixth Joint Swedish-Russian Int. Workshop on Inform. Theory*, Mölle (1993) 362-366.

[106] T. Kløve, "Bounds on the weight distribution of cosets", manuscript 1993.

[107] T. Kløve, "On Massey's bound on the worst-case probability of undetected error", *Proc. 1994 IEEE Int. Symp. Inform. Theory*, (1994) 242.

[108] T. Kløve, "Bounds on the worst-case probability of undetected error", *IEEE Trans. Inf. Theory*, 41 (1995) 298-300.

[109] T. Kløve, "Reed-Muller codes for error detection", manuscript 1994.

[110] T. Kløve, "The worst-case probability of undetected error for linear codes on the local binomial channel", to appear in *IEEE Trans. Inform. Theory.*

[111] T. Kløve, "Near-MDS codes for error detection", *Proc. Int. Workshop on Optimal Codes and Related Topics*, Sozopol, Bulgaria, 26 May-1 June 1995, 103-107.

[112] T. Kløve, "Almost MDS and Near-MDS codes for error detection", to appear in *IEEE Trans. Inform. Theory.*

[113] T. Kløve and M. Miller, "The detection of error after error-correction decoding", *IEEE Trans. on Comm.* 32 (1984) 511-517.

[114] E. Kolev and N. Manev, "The binary weight distribution of an extended $(2M, 5)$ Reed-Solomon code and its dual", *Doklady Bolgarskoi Akademii Nauk* 43, Iss. 9 (1990) 9-12.

[115] E. Kolev and N. Manev, "The binary weight distribution of the extended $[2^m, 5]$ Reed-Solomon code and its dual code" (in Russian), *Prob. Peredach. Inform.* 30, no. 3 (1994) (in Russian) [English translation: *Prob. Inform. Transmission*].

[116] V. I. Korzhik, "Bounds on undetected error probability and optimum group codes in a channel with feedback", *Radioteckhnika* 20, vol. 1 (1965) 27-33 (in Russian) [English translation: *Telecommun. Radio Eng.* 20, no.1, part 2 (1965) 87-92].

[117] V. I. Korzhik and S. D. Dzubanov, "Error detection codes for local binomial channels", *Proc. The Third Int. Colloquium on Coding Theory*, Dilijan, Armenia, Sept. 25 - Okt. 2 (1990) 127-130.

[118] V. I. Korzhik and L. M. Fink, "The multistage stochastic coding", *Int. Symp. Inform. Theory*, Repino, USSR (1976).

[119] V. I. Korzhik, M. Y. Lesman and P. Malev, "An ARQ scheme using concatenated codes for mobile satellite channels and its throughput efficiency optimization", *Proc. Int. Conf. Russat '93*, St. Petersburg (1993) 39-42.

[120] V. I. Korzhik, S. A. Osmolovskii, and L. M. Fink, "Universal stochastic coding in decision-feedback systems", *Prob. Peredachi Inform.* 10, no. 4 (1974) 25-29 (in Russian) [English translation: *Prob. Inform. Transmission* 10 (1974) 296-299].

[121] V. I. Korzhik and V. A. Yakovlev, "Nonasymptotic estimates of information protection efficiency for the wire-tap channel concept", Auscrypt 92, *Springer Lecture Notes in Computer Science* 718 (1992) 185-195.

[122] M. A. Kousa and M. Rahman, "An adaptive error control-system using hybrid ARQ schemes", *IEEE Trans. Commun.* 39 (1991) 1049-1057.

[123] H. Krishna and S. D. Morgera, "A new error control scheme for hybrid ARQ systems", *IEEE Trans. on Comm.* 35 (1987) 981-990.

[124] P. V. Kumar and V. Wei, "Minimum distance of logarithmic and fractional partial m-sequences", *IEEE Trans. Inform. Theory* 38 (1992) 1474-1482.

[125] S. Kundu and S. M. Reddy, "On symmetric error correcting and all unidirectional error detecting codes", *IEEE Int. Symp. Inform. Theory* (1988) 120.

[126] A. Kuznetsov, F. Swarts and H. C. Ferreira, "On the undetected error probability of linear block codes on channels with memory", *Proc. Sixth Joint Swedish-Russian Int. Workshop on Inform. Theory*, Mölle (1993) 367-371.

[127] A. Kuznetsov, F. Swarts, A. J. Han Vinck and H. C. Ferreira, "On the undetected error probability of linear block codes on channels with memory", Manuscript, September 1993.

[128] J. Lawson, A. Michelson, and Z. Huntoon, "A performance analysis of a family of codes constructed from simple binary codes", presented at *Canadian Conf. Commun. and Power*, Montreal, PQ, Canada, Oct. 1978.

[129] V. K. Leontev, "Asymptotic optimum group codes in channels with feedback", presented at *All-Union Conf. on Probl. of Theoretical Cybernetics*, Novosibirsk 1969 (in Russian).

[130] V. K. Leontev, "Error-detecting encoding", *Prob. Peredach. Inform.* 8, no. 2 (1972) 6-14 (in Russian) [English translation: *Prob. Inform. Transmission* 8 (1972) 86-92].

[131] S. K. Leung, "Evaluation of the undetected error probability of single parity-check product codes", *IEEE Trans. Commun.* 31 (1983) 250-253.

[132] S. K. Leung and M. E. Hellman, "Concerning a bound on undetected error probability", *IEEE Trans. Inform. Theory* 22 (1976) 235-237.

[133] S. K. Leung, E. R. Barnes, and D. U. Friedman, "On some properties of the undetected error probability of linear codes", *IEEE Trans. Inform. Theory* 25 (1979) 110-112.

[134] C. Leung and K. A. Witzke, "On testing for improper error detection codes", *IEEE Trans. Commun.* 38 (1990) 2085-2086.

[135] V. I. Levenshtein, "Bounds on the probability of undetected error", *Prob. Peredachi Inform.* 13, no.1 (1977) 3-18 (in Russian) [English translation: *Prob. Inform. Transmission* 13 (1978) 1-12].

[136] V. I. Levenshtein, "Straight-line bound for the undetected error exponent", *Prob. Peredachi Inform.* 25 no.1 (1989) 33-37 (in Russian) [English translation: *Prob. Inform. Transmission* 25 (1989) 24-27].

[137] M.-C. Lin, "Bounds on the undetected error probabilities of linear codes for single-error correction and error detection", *Electronic Letters* 27 (1991) 2264-2265.

[138] M.-C. Lin, "Undetected error probabilities of codes for both error correction and detection", *IEEE Trans. Inform. Theory* 36 (1990) 1139-1141.

[139] M. Ch. Lin and M. Y. Gun, "The performance analysis of a concatenated ARQ scheme using parity retransmissions", *IEEE Trans. Commun.* 39 (1991) 1869-1874.

[140] D.-L. Lu and J.-F. Chang, "The effect of return channel errors on the performance of ARQ protocols", *Proc. 1991 IEEE Int. Symp. Inform. Theory* (1991) 221.

[141] D.-L. Lu and J.-F. Chang, "Performance of ARQ protocols in bursty channel errors", *Proc. 1994 IEEE Int. Symp. Inform. Theory*, (1994) 375.

[142] J. MacWilliams, "A theorem on the distribution of weights in a systematic code", *Bell Syst. Tech. J.* 42 (Jan. 1963) 79-94.

[143] P. Malev, *Investigation and Design of Effective Methods for Digital Communication with ARQ* (in Russian), PhD Thesis, LEIC, St. Petersburg, Russia (1993).

[144] C. L. Mallow, V. Pless, and N. J. A. Sloane, "Self-dual codes over GF(3)", *SIAM J. Applied Math.* 31 (1976) 649-666.

[145] J. Massey, "Coding techniques for digital data networks" in *Proc. Int. Conf. Inform. Theory Syst.*, NTG-Fachbeirichte 65, Berlin, Germany, Sept. 18-20 1978, 307-315.

[146] R. J. McEliece and H. Rumsey Jr., "Euler products, cyclotomy, and coding", J. Number Theory 4 (1972) 302-311.

[147] A. T. Memişoğlu and S. Bilgenr, "An adaptive hybrid ARQ scheme based on exponentially weighted moving averages for slowly time-varying communication channels", *Proc. 1994 IEEE Int. Symp. Inform. Theory*, (1994) 377.

[148] M. J. Miller, "The probability of undetected error for shortened cyclic codes", *IEEE Int. Symp. Inform. Theory* (1985) 111.

[149] M. J. Miller and S. Lin, "Codes for error detection in data transmission and storage", *J. of Electrical and Electronics Eng., Australia* 3 (Sept 1986).

[150] M. J. Miller, M. G. Wheal, G. J. Stevens, and A. Mezhvinsky, "The X.25 error detection code is a poor choice" *Proc. of the Institution of radio and Electronic Engineers Australia IREECON '85*, Sydney, September 1985.

[151] M. J. Miller, M. G. Wheal, G. J. Stevens, and S. Lin, "The reliability of error detection schemes in data transmission", *J. of Electrical and Electronics Eng., Australia* 6 (June 1986) 123-131.

[152] T. Nishijima, O. Nagata, and S. Hirasawa, "On the probability of undetected error for iterated codes", *Proc. 1993 IEEE Int. Symp. Inform. Theory*, (1993) 162.

[153] T. Nishijima and S. Hirasawa, "On the probability of undetected error and the computational complexity to detect an error for iterated codes", to be presented at the 1995 IEEE Int. Symp. Inform. Theory.

[154] V. K. Oduol and S. D. Morgera, "Performance evaluation of the generalized type-II hybrid ARQ scheme with noisy feedback on Markov channels", *IEEE Trans. Commun.*, 41 (1993) 32-40.

[155] C. T. Ong and C. Leung, "On the undetected error probability of tripel-error-correcting BCH codes", *IEEE Trans. Inform. Theory* 37 (1991) 673-678.

[156] R. Padovani and J. K. Wolf, "Data transmission using error detection codes", in GLOBECOM '82, IEEE Global Telecom. Conf., Miami, Nov. 29-Dec. 2, 1982, 626-631.

[157] R. Padovani and J. K. Wolf, "Poor error correction codes are poor error detection codes", *IEEE Trans. Inform. Theory* 30 (1984) 110-111.

[158] P. Perry, "Necessary conditions for good error detection", *IEEE Trans. Inform. Theory* 37 (1991) 375-378.

[159] P. Perry, "Runlength-limited codes for single error detection in magnetic recording channel", *IEEE Trans. Inform. Theory* 41 (1995) 809-815.

[160] D. K. Pradhan, "A new class of error-correcting/detecting codes for fault-tolerant computer applications", *IEEE Trans. Comput.* 29 (1980) 471-481.

[161] D. K. Pradhan and S. K. Gupta, "A new framework for designing and analyzing BIST techniques and zero aliasing compression", *IEEE Trans. Comput.* 40 (1991) 743-763.

[162] D. K. Pradhan, S. K. Gupta, and M. G. Karpovsky, "Aliasing probability for multiple input signature analyzer", *IEEE Trans. Comput.* 39 (1990) 586-591.

[163] I. Rahman, "Nonbinary error detection codes for data retransmission and bit error rate monitoring", *IEEE Trans. Commun.* 40 (1992) 1139-1143.

[164] L. K. Rasmussen and S. B. Wicker, "The performance of type-I, trellis coded hybrid-ARQ protocols over AWGN and slowly fading channels", *IEEE Trans. Inform. Theory*, to appear.

[165] L. K. Rasmussen and S. B. Wicker, "Trellis coded, type-I hybrid-ARQ protocols based on CRC error-detecting codes", *IEEE Trans. Commun.* 40 (1994) 418-428.

[166] L. K. Rasmussen and S. B. Wicker, "A comparison of two code combining techniques for equal gain, trellis coded diversity receivers", *IEEE Trans. on Vehicular Tech.* 44 (1995) 291-295.

[167] G. R. Redinbo, "Inequalities between the probability of a subspace and the probabilities of its cosets", *IEEE Trans. Inform. Theory* 19 (1973) 533-36.

[168] C. Retter, "The average binary weight-enumerator for a class of generalized Reed-Solomon codes", *IEEE Trans. Inform. Theory* 37 (1991) 346-349.

[169] C. Retter, "Gaps in the binary weight distribution of Reed-Solomon codes", *IEEE Trans. Inform. Theory* 38 (1992) 1688-1697.

[170] M. Rice and S. Wicker, "Modified majority logic decoding of cyclic codes in hybrid-ARQ systems", *IEEE Trans. Commmun.* 40 (1992) 1413-1417.

[171] John Riordan and N.J.A Sloane, "The enumeration of rooted trees by total height" *J. Australian Math. Soc.* 10 (1969) 278-82.

[172] F. Rodier, "On the spectra of the duals of binary BCH codes of designed distance $\delta = 9$", *IEEE Trans. Inform. Theory* 38 (1992) 478-479.

[173] R. Schoof and M. van der Vlugt, "Hecke operators and the weight distribution of certain codes", *J. Combin. Theory Ser. A* 57 (1991) 163-186.

[174] J. W. Schwartz and J. K. Wolf, "A systematic (12,8) code for correcting single errors and detecting adjacent errors", *IEEE Trans. Comput.* 39 (1990) 1403-1404.

[175] R. Segal and R. Ward, "Weight distributions of some irreducible cyclic codes", *Mathematic of Computation* 46 (1986) 341-354.

[176] N. Seshadri and C.-E. W. Sundberg, "List Viterbi decoding algorithms with applications", *IEEE Trans. Commmun.* 42 (1994) 313-323.

[177] C. E. Shannon, "A mathematical theory of communication", *Bell Syst. Tech. J.* 27 (1948) 379-423 & 623-656.

[178] W. M. Sidelnikov, "Weight spectrum of binary Bose-Chaudhuri-Hocquenghem codes" (in Russian), *Prob. Peredachi Inform.* 7, no. 1 (1971) 14-22. [English translation: *Prob. Inform. Transmission* 7 (1971) 11-17].

[179] K. N. Sivarajan, R. McEliece, and H. C. A. van Tilborg, "Burst-error-correcting and detecting codes", *IEEE Int. Symp. Inform. Theory* (1990) 153.

[180] N. J. A. Sloane and E. R. Berlekamp, "Weight enumerator for second-order Reed-Muller codes", *IEEE Trans. Inform. Theory* 16 (1970) 745-751.

[181] P. Solé, "A limit law on the distance distribution of binary codes", *IEEE Trans. Inform. Theory* 36 (1990) 229-232.

[182] R. C. Sommer, "A note upon another error detecting code that is not proper", *IEEE Trans. Commun.*, 33 (1985) 719-721.

[183] D. D. Sullivan, "A fundamental inequality between the probabilities of binary subgroups and codes", *IEEE Trans. Inform. Theory* 13 (1967) 91-94.

[184] L. Swanson and R. McEliece, "Reed-Solomon codes for error detection", *IEEE Int. Symp. Inform. Theory* (1985) 110.

[185] F. Swarts, A. J. Han Vinck, and H. C. Ferreira, "Undetected error probability of linear block codes on channels with memory", *Proc. 1993 IEEE Int. Symp. Inform. Theory*, (1993) 162.

[186] W. Szpankowski, "On asymptotics of certain sums arising in coding theory", to appear in *IEEE Trans. Inform. Theory.*

[187] N. Tanaka and S. Kaneki, "A class of codes which detect deception and its application to data security", *Proc. Carnahan Conf. on Crime Countermeasures*, April 6-8, 1977, Kentucky (1977) 221-227.

[188] A. I. Turkin and V. I. Korzhik, "A practical optimal decoding algorithm for arbitrary linear codes having polynomial complexity", Presentation at *1991 IEEE Int. Symp. Inform. Theory.*

[189] G. van der Geer and M. van der Vlugt, "Weight distributions for a certain class of codes and maximal curves", *Discrete Math.* 106 (1992) 209-218.

[190] G. van der Geer, R. Schoof, and M. van der Vlugt, "Weight formulas for ternary Melas codes", *Mathematics of Computation* 58 (1992) 781-792.

[191] N. H. Vaiday and D. K. Pradhan, "A new class of bit- and byte-error control codes", *IEEE Trans. Inform. Theory* 38 (1992) 1617-1623.

[192] X. Wang, "The existence and conjecture of binary $(N, 2, \omega)$ optimal error-detecting constant weight codes", *Scientia Sinica Series A - Mathematical Physical Astronomical & Technical Sciences* 31, iss.5 (1988) 621-629.

[193] X. Wang, "The undetected error probability of binary (2m,2,m) nonlinear constant weight codes", manuscript 1993.

[194] X. Wang and Y. Yang, "On the undetected error probability of nonlinear binary constant weight codes", *IEEE Trans. Commun.* 42 (1994) 2390-2394.

[195] R. K. Ward and M. Tabaneh, "Error correction and detection, a geometric approach", *The Computer Journal* 27 no. 3 (1984) 246-253.

[196] R. L. Ward, "Weight enumerators of more irreducible cyclic binary codes", *IEEE Trans. Inform. Theory* 39 (1993) 1701-1709.

[197] G.N. Watson, "Theorems stated by Ramanujan (V): Approximations connected with e^x" *Proc. London Math. Soc* (2) 29 (1929) 293-308.

[198] J. H. Weber, C. de Vroedt, and D. E. Boekee, "On codes correcting/detecting symmetric, unidirectional, and/or asymmetric errors", *IEEE Int. Symp. Inform. Theory* (1990) 86.

[199] M. N. Wegman and J. L. Carter, "New hash functions and their use in authentication and set equality", *J. Comp. System Sci.* 22 (1981) 265-279.

[200] V. K. Wei, "Generalized Hamming Weights for Linear Codes", *IEEE Trans. Inform. Theory* 37 (1991) 1412-1418.

[201] S. B. Wicker, "Reed-Solomon error control coding for Rayleigh fading channels with feedback", *IEEE Trans. on Vehicular Tech.* 41 (1992) 124-133.

[202] S. B. Wicker and M. Bartz, "Type-II hybrid-ARQ protocols using punctured MDS codes", *IEEE Trans. Commun.* 42 (1994) 1431-1440.

[203] S. B. Wicker and M. Bartz, "The design and implementation of type-I and type-II hybrid-ARQ protocols based on first-order Reed-Muller codes", *IEEE Trans. Commun.*, 42 (1994) 979-987.

[204] K. A. Witzke, "Examination of the undetected error probability of linear block codes", M.A.Sc. Thesis, Dep. Elec. Eng., Univ. British Columbia, Vancouver, Canada (1984).

[205] K. A. Witzke and C. Leung, "A comparison of some error detecting CRC code standards", *IEEE Trans. Commun.* 33 (1985) 996-998.

[206] J. K. Wolf and R. D. Blakeney II, "An exact evaluation of the probability of undetected error for certain shortened binary CRC codes", *IEEE MILCOM 88,* (1988) 287-292.

[207] J. K. Wolf and D. Chun, "The single burst error detection performance of binary cyclic codes", *IEEE Trans. Commun.* 42 (1994) 11-13.

[208] J. K. Wolf, A. M. Michelson, and A. H. Levesque, "The determination of the probability of undetected error for linear binary block codes", in *Int. Conf. on Commun.*, Denver, Colorado, June 14-18, 1981, 65.1.1-5.

[209] J. K. Wolf, A. M. Michelson, and A. H. Levesque, "On the probability of undetected error for linear block codes", *IEEE Trans. Commun.* 30 (1982) 317-324.

[210] A. Wyner, "The wire-tap channel", *Bell Syst. Tech. J.* 54 (1975) 1355-1381.

[211] D. Xu, "A dual theorem on error detection codes" (in Chinese), *Journal of Nanjing Aeronautical Inst.* 24 No. 6 (1992) 730-735.

[212] H. Yamamoto and K. Itoh, "Viterbi decoding algorithm for convolutional codes with repeat request", *IEEE Trans. Inform. Theory* 26 (1980) 540-547.

[213] V. V. Zyablov, "The performance of the error correcting capability of the iterative and concatenated codes" (in Russian), In: *Digital Information Transmission over Channels with Memory*, Nauka, Moscow 1970.

INDEX

NOTATIONS

MIX
Papier aus verantwortungsvollen Quellen
Paper from responsible sources
FSC® C105338

If you have any concerns about our products,
you can contact us on
ProductSafety@springernature.com

In case Publisher is established outside the EU,
the EU authorized representative is:
Springer Nature Customer Service Center GmbH
Europaplatz 3, 69115 Heidelberg, Germany

Printed by Libri Plureos GmbH
in Hamburg, Germany